Retail Location Planning in an Era of Multi-Channel Growth

The way in which products and services are delivered to consumers, through branches and retail outlets, or more generally through a network of distribution channels, remains fundamentally important for maintaining a competitive advantage for a very wide range of businesses. This is true within domestic markets, but especially so for increasingly global corporations, as shareholder pressure for continued growth drives businesses into ever more widespread geographical markets.

Arguing that more complex markets demand more sophisticated spatial analysis, this book discusses the application of location planning techniques to generate competitive advantage in a variety of business sectors in a changing retail environment. The series of techniques are analysed, from relatively straightforward branch scorecards to sophisticated applications of geographical information systems (GIS), spatial modelling and mathematical optimisation. Also explored are the changing dynamics of the impact of more restrictive planning environments in many countries on how retailers find new locations for growth and respond to changing consumer needs and wants.

The book is essential reading for students and scholars alike working in geography, economics, business management, planning, finance and industry studies.

Mark Birkin is Professor of Spatial Analysis and Policy and Director of the Consumer Data Research Centre (CDRC) at the University of Leeds. His major interests are in simulating social and demographic change within cities and regions, and in understanding the impact of these changes on the need for services like housing, roads and hospitals, using techniques of microsimulation, agent-based modelling and GIS. He has also been involved with many retail-based projects with a number of major blue-chip clients.

Graham Clarke is Professor of Business Geography at the University of Leeds. He has worked extensively in various areas of GIS and applied spatial modelling, focusing on many applications within urban/social geography.

Graham specialises in retail geography and model development in relation to retail store location planning. His major research interests relate to retail location planning in relation to the multi-channel growth strategies of retail organisations.

Martin Clarke is Professor of Geographic Modelling in the School of Geography, University of Leeds, and Deputy Director of the Consumer Data Research Centre. Martin's main interests are based around service analysis and planning. From 1990 to 2004 he was Chief Executive of GMAP Ltd, one of the most successful university spin-out companies in the UK that specialised in network planning and location analysis for some of the world's largest retail corporations.

Retail Location Planning in an Era of Multi-Channel Growth

Mark Birkin, Graham Clarke and Martin Clarke

Routledge
Taylor & Francis Group
LONDON AND NEW YORK

First published 2017 by Routledge

2 Park Square, Milton Park, Abingdon, Oxfordshire OX14 4RN
52 Vanderbilt Avenue, New York, NY 10017

Routledge is an imprint of the Taylor & Francis Group, an informa business

First issued in paperback 2018

British Library Cataloguing in Publication Data
A catalogue record for this book is available from the British Library

Library of Congress Cataloging in Publication Data
Names: Birkin, Mark, author | Clarke, Graham, 1960– author. |
Clarke, M. (Martin), 1955– author.
Title: Retail location planning in an era of multi-channel growth /
Mark Birkin, Graham Clarke and Martin Clarke.
Description: Abingdon, Oxon; New York, NY: Routledge, 2017.
Identifiers: LCCN 2016047492 | ISBN 9781409404071 (hardback) |
ISBN 9781315605937 (ebook) Subjects: LCSH: Store location–Planning. |
Stores, Retail–Planning. | Retail trade.
Classification: LCC HF5429.275 .B573 2017 | DDC 658.8/7–dc23
LC record available at https://lccn.loc.gov/2016047492

ISBN: 978-1-4094-0407-1 (hbk)
ISBN: 978-0-367-21875-1 (pbk)

Typeset in Times New Roman
by Out of House Publishing

Contents

Figures

Tables

Acknowledgements

The authors are very grateful to all the publishers and individuals who have kindly granted permissions to use their figures or tables in this book. We are especially grateful to Nick Henthorn (Telefonica Data Insights); Jesse Pearson (Office of Transportation System Management (OTSM) Minnesota Department of Transportation; John Stillwell (University of Leeds); Martin Bradbury (Callcredit Information Group); Andy Newing (University of Leeds); Pakorn Meksangsouy (former PhD student at the University of Leeds); and Heather Eyre/Ross (former PhD student at the University of Leeds). Alison Manson (University of Leeds) also kindly drew a number of the figures from scratch.

We would also like to thank the various editorial managers/assistants who have helped us to put the book together, in particular Pris Corbett, Emma Chappell, Cathy Hurren and Heidi Cormode. In addition, Chris Steel worked tirelessly to copy-edit and standardise the entire book in preparation for the proofs – thanks for a great job, Chris.

Finally, our thanks go to the many colleagues and students who, over the years, have provided feedback on our work and suggestions for new avenues of research.

1 Introduction

The central message of this book is that the way in which products and services are delivered to consumers, through branches and retail outlets, or more generally through a network of distribution channels, remains fundamentally important for maintaining a competitive advantage for a very wide range of businesses. This is true within domestic markets, but especially so for increasingly global corporations, as shareholder pressure for continued growth drives businesses such as Walmart, IKEA, General Motors, Burger King (to name but a few) into ever more widespread geographical markets.

In this book we will discuss the application of location planning techniques to generate competitive advantage in a variety of business sectors, building on, and updating, the framework introduced in Birkin, Clarke and Clarke (2002). The array of techniques that we will describe ranges from relatively straightforward branch scorecards to sophisticated applications of geographical information systems (GIS), spatial modelling and mathematical optimisation. We shall also place these techniques within the changing retail environment, especially in relation to the UK. With more restrictive planning environments in many countries, retailers are becoming cleverer at finding new locations for growth and responding to changing consumer needs and wants. An exploration of these dynamics is also a major focus of the book.

An important innovation in retail location over the last 20 years or so has been the availability of better data and information. Retail and service organisations now know more than ever before about their customers and their behaviour, and about the performance of their own distribution networks. For example, a retail organisation may use electronic point-of-sale systems to monitor stocking levels on its shelves and in its warehouses during specific times of each day. Those businesses which employ loyalty cards may not only track the source of customer spending within their stores, but also purchase frequency, basket size and (by implication) share of wallet.

The ideas and methodologies we present can be used in a wide range of business situations to generate important benefits, both immediately to the bottom-line and to longer-term business strategy. Thus location planning techniques may be used to benchmark the performance of existing branches or networks; to evaluate accurately 'what-if?' scenarios involving new store builds or network reconfiguration; to optimise distribution as part of a long-term strategy or entry to a new market; and may even provide great benefits within the merger and acquisition process.

These ideas may seem counterintuitive in the modern retail environment to the extent that many commentators and practitioners have tended to assume that ongoing changes in modes of communication and service delivery, particularly the internet and online retailing, have begun to spell what Cairncross (1997) first referred to as the 'death of distance'. In the book we will demonstrate that this is very far from the case. In typical situations, we will argue that for the majority of businesses what this really means is that the business environment through which customer needs are satisfied is actually more complicated and difficult to understand than ever before. The widespread failure of many internet pure-play retail businesses (such as Webvan and Virgin Cars) seems to endorse the argument that, more often than not, the importance of e-commerce lies in its ability to complement and not replace existing channels.

An excellent example of the complexity of contemporary distribution channel networks can be found in the financial services industry. Historically, the only touch-point between the customer and the organisation might have been through the branch and its local manager. Now customers will expect to be able to withdraw cash or query accounts through a cash-point machine (ATM), to find product information or current rates on the internet, to obtain 24/7 advice and guidance by telephone, and to have stock quotes or overdraft alerts sent directly by text or email. They might want to make a loan application via interactive television. Furthermore, they will expect to play off one provider against another not just for different financial service products, but for the same product on different occasions – thus multiple cheque accounts, credit cards, stock dealing advisers and so forth.

Retail location planning remains important for a number of reasons. One major consideration is that all organisations experience huge variations in branch performance and profitability. For every award winning motor sales franchise, there is another who is allowing competing manufacturers to dominate the local market. Every bank has ATMs which are used infrequently and every supermarket retailer has stores which will never generate a return on the capital employed in their development. Furthermore, as we have stated already, distribution options are increasingly complex. Retail and service businesses need to meet the needs of their customers throughout the week, often involving service provision not just at home, but from the workplace, while on holiday or at school, at the theatre, restaurant or golf course, or while in transit between these various activities. And so, for example, the configuration of new vehicle sales franchises may need to be very different from aftersales: the first of these dictated by an important long-term capital investment which might be influenced by the whole family; the latter a distress activity where inconvenience needs to be minimised. Similarly, the opportunity for petrol companies to maximise returns from forecourts will be very different on motorways or interstate highways than in local residential neighbourhoods.

Another important consideration is that organisations spend vast amounts on their physical distribution channels. For example, in 2015 Walmart had over 11,000 stores in 28 countries, employing 2.2 million persons in total. The process of retail globalisation has characteristically been driven through this physical expansion of retail networks, rather than by their virtual extension through e-commerce. The British supermarket retailer Tesco, for example, has grown into a £70 billion corporation (2015 figures) mainly through expansion of its store portfolios: first UK hypermarkets and then into Eastern Europe and the Far East. Although it is now also the world's most important grocery internet player, internet retailing still accounts for a small percentage of Tesco's sales worldwide.

From what we have said above, it is clear that location planning is an activity with profound practical and real world importance. We also note that location questions impact on many business functions, including operations management, strategic planning, marketing, sales, property, finance and logistics. Any effective business organisation that is engaged in the delivery of products and services will be continually reviewing its sales channels at both an operational and a strategic level. This book will discuss examples by which spatial analysis methods can be used to predict (accurately!) the sales turnover of a new outlet and its impact on existing outlets. If this can be done then retailers can:

- evaluate new potential sites to exploit greater sales;
- direct expenditure on local advertising and promotion;
- evaluate the fit between two organisations which plan to merge and also to provide insights on branding, network consolidation and other issues post-merger; plan ideal sales territories for franchise retailers;
- benchmark whether weak profitability in an existing location or region is caused by poor performance or adverse market conditions;
- find the ideal places in a network to promote a new product or service; and find the right way to benefit from a new or enhanced method of delivery for an existing product or service.

Despite the more competitive and restrictive planning environment in many countries it is interesting that not all retailers use the range of sophisticated techniques now available. Indeed, we argue that many senior executives in retailing have historically underestimated the importance of location planning. Going back as far as the 1960s, most retailers relied on 'gut feeling' and 'checklist' approaches to evaluate the potential performance of selected retail sites. Gut feeling is usually thought of as the simplest in terms of spatial analysis. It normally involves the on-site decision of a senior member of staff who obtains a 'gut feeling' for a location through a site visit. Perhaps this reflects a lack of location planning modules in business and management schools (generally) which would help to increase the recognition of the need for geographical space within the cycles of production and consumption. For example, how many MBA courses include modules on spatial planning? The lack of computer models used in site location has also been picked up in a number of articles which have interviewed key staff and found a mixture of apathy towards, and mistrust of, so-called sophisticated methods. This was probably first discussed by Hernandez and Bennison (2000) and I. Clarke *et al.* (2003). More recently an interesting collection of papers by Reynolds and Wood (2010) and Wood and Reynolds (2011; 2012) has reaffirmed that a lot of (even large) firms rely more on senior experience. Many of the location teams in the companies interviewed by Wood and Reynolds argue that their power to influence decision making is very limited – at best often providing simple maps and data to the more powerful marketing teams.

Of course we would never deny the importance of experience. Even those analysts that use computer models on a regular basis will spend a lot of time out of the office, checking the site characteristics, the nature of the competition, access etc. They sensibly use the computer models as decision support systems – one part of the jigsaw necessary to make intelligent decisions. Those that do use such methodologies generally seem more linked into the actual store location decision-making process – taking responsibility for store turnover forecasts and feeling much more equal with marketing and finance departments (Wood and Reynolds 2011; 2012). Perhaps these retailers understand that a strategy based largely on experience alone has its own obvious problems. Will senior executives always outperform technology? For all the success stories given by senior executives in relation to finding good sites using their intuition or experience, we could give examples where other individuals have got it badly wrong. Such mistakes are rarely admitted, however.

In addition to students of retail geography, marketing and management we hope that this book will have widespread appeal to managers in organisations which are engaged in service delivery to a spatially distributed customer base. The concepts to be discussed are of interest and importance to both large and small organisations, and to managers in many different functional divisions of those organisations. In particular, we will seek to demonstrate that distribution channel planning and management is not an activity that can be compartmentalised, but is fundamentally bound up with both product planning and marketing through a triangle of interactions between customers, channels and products (see Figure 1.1).

In addition to the many organisations that have a direct interest in product or service delivery, there are others with an indirect interest. For example, brand managers within manufacturing organisations have a strong interest in product placement within partner retail organisations, even though they may not have direct sales channels of their own. Moreover it is worth noting that all managers are also consumers of products and services, and so from this perspective there is potential interest for the whole readership, albeit at a personal rather than a professional level.

For the last 30 years or so the authors have been engaged in geographical modelling and spatial analysis as both academics and practitioners. Through our professional work in particular, we have worked with many organisations

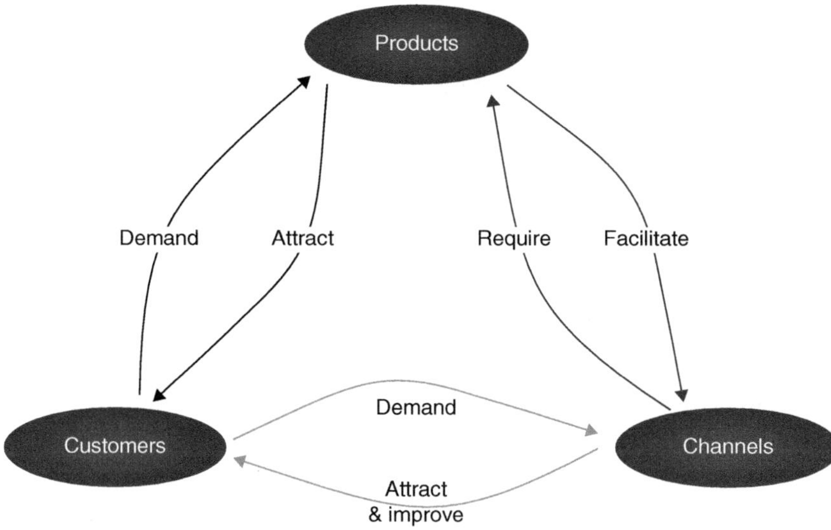

Figure 1.1 The triangle of distribution.
Source: Authors

to develop 'spatial decision support systems' which inform the business planning process at all levels from operations to strategy. These spatial decision support systems combine empirical data about customer spending and behaviour, about outlets and business performance, about infrastructure, competition, regulation and economic markets. Thus the inferences made from the modelling and spatial analysis activities which we describe are thoroughly grounded in the ability to understand and reproduce real customer behaviours and business performance. Organisations such as Ford and HBoS have fundamentally changed their distribution systems and retail structure in response to this analysis. For example, Ford has adopted consistent branding and ownership within newly designed 'customer marketing areas', with major increases in competitiveness and network efficiency (see Chapter 8 for more details).

The rest of the book is organised as follows. In Chapter 2 we explore some new issues that are impacting generally on store location planning as background to the material which follows. This includes the concept of niche spatial marketing. Greater planning restrictions on traditional sites (especially large out-of-town developments) in many countries are forcing retailers to rethink new types of location to exploit, making the contemporary store location planning agenda more diverse than ever before. In the next few chapters we explore retail store location methodologies in more detail. In Chapter 3 we look at the growing use of GIS for retail location planning. We provide a critical review of strengths and weaknesses. As discussed, GIS is often synonymous in retail analysis with geodemographics. Hence we review the use of geodemographics in retail planning in Chapter 4. In Chapter 5 we introduce a broad review of other techniques for site location analysis – including analogue, regression and spatial interaction or gravity models.

In the next two chapters we explore in more depth the two most important ingredients in store location methodologies – demand and supply. First, in Chapter 6, we undertake a broad review of retail demand. Here we look at traditional ways of measuring demand (for input into whichever site location methodology is preferred) and how demand is changing, both spatially and in terms of socio-economic and lifestyle changes. In Chapter 7 we consider the supply-side characteristics of retail environments – in particular store or site attractiveness. A key question remains what are the most important drivers of attractiveness to different types of retail destination?

In Chapter 8 we review progress with optimisation in retail modelling approaches. This is followed by a consideration of network reinvention in Chapter 9: that is, the use of spatial models and analysis to help plan the redesign of entire networks which perhaps were invented many years ago and no longer seem fit for purpose. In the final chapters we switch focus to look at how developments in consumer use of technology might impact on store location in the future. First in Chapter 10 we look at why geography is important for e-commerce (despite Caincross's pessimistic scenario mentioned above) and how retailers may be able to plan their combined 'click and brick' strategy more effectively in the future. Finally we explore how big data, driven by the increased use of mobile phones and social media, has the potential to help site location teams. Again we will draw mainly on our own research work undertaken in the Economic and Social Research Council (ESRC) 'Consumer Data Research Centre' set up at the University of Leeds in 2014. This multi-million pound investment by the UK government intends to exploit and build on the links between academia and commercial organisations by trying to make commercial data sets more available to the academic community. This future collaborative project builds on the success of many previous academia/retail business linkages at the University of Leeds, many examples of which are given throughout the book.

2 The dynamics of retail store location

2.1 Introduction

Alongside many demand-side changes (see Chapter 6), many retailers are feeling the pressure of growing competition, stricter planning legislation and the perceptions of increasing saturation in many markets. The combined effects of these dynamics are that retailers are continually looking for new ways to grow. This has been especially prominent in Europe since the mid to late 1990s when many governments introduced stricter planning guidelines towards retail location policy, especially in relation to large stand-alone edge of town formats. The aim of this chapter is to explore the pressures on traditional growth models and show how retailers are reacting in part by searching for new retail spaces – niche spatial marketing – using new retail formats. This, in turn, presents new challenges for store location teams as the retail environment gets more sophisticated in terms of the number of retail channels, formats and the types of location they must plan to deal with. In Section 2.2 we explore how traditional growth models are under threat from a variety of factors and briefly consider the impact on store location research. In Section 2.3 we start to look at a number of reactions to these changes. The first of these is the growth of the discount market (in various formats) and the search for new locations to maximise sales. In Section 2.4 we explore the growth of the convenience market from the supply-side (demand-side changes given in Chapter 6) and again the search for new locations. In Section 2.5 we discuss a more miscellaneous set of location issues – new opportunities provided by transport hubs, (re)considerations of tourism spaces and the drive to international markets.

2.2 Pressures on traditional growth models

In many developed world markets, the late 1970s, the 1980s and the early 1990s proved to be a golden era for retail growth. This was true across much of Western Europe, USA, Canada, Japan and Australia. Although planning restrictions were in place in many of these countries (Guy 1998 and Howe 2003 look at similarities and differences across Europe) such legislation was generally ineffective in stopping the tide of out-of-town developments that took place (Burt 1984 and Cliquet 2000, for example, both give explanations for the continued growth of French superstores despite the greater legislation against stand-alone superstores ['Loi Royer'] introduced in 1973). In some countries, such as the UK and later the USA, the government's laissez-faire attitude to planning (especially under the Thatcher and Reagan regimes respectively) actually encouraged an era of large-scale superstore and regional shopping centre growth. In the grocery market, the 'store wars' battle that ensued (Wrigley 1987) was all about obtaining the best sites for large superstores and hypermarkets, often in green-field locations at the edge of more affluent suburbs of towns and cities. This also encouraged many of the larger firms to invest heavily in store location and GIS as the more straightforward site location models associated with town centres became inappropriate to use for new green-field sites.

The story of growth and change varies from country to country and we focus on the UK in detail below. We hope non-UK readers will recognise similar changes in their own countries and could put together a comparative narrative (for example, see the useful account of the geography of retail change in Sweden by Amcoff 2016). However, where possible we do make some comparisons to developments outside the UK.

The laissez-faire UK retail environment began to change in the 1990s (Wrigley 1991, 1994, 1998). First, new competition arrived in the UK in the form of the deep discounters from Scandinavia and Germany (this was certainly mirrored in other parts of Europe: see Colla 2003, 2004; Poole *et al.* 2002b; Wortmann 2004). Second, the leading grocery multiples were forced to depreciate their assets in the early 1990s as it was widely believed that retailers had paid too much for land, and they would not get the price they paid for that land if they were forced to sell, especially for other land-uses (i.e. assets were overvalued on company balance sheets). Third, new planning legislation was introduced in 1996 in the UK – PPG6, designed to protect town and city centres in the future and stop the spread of yet more green-field out-of-town developments. This was mirrored in other parts of Europe where similar legislation such as France's 'Loi Raffarin' in 1996 was introduced. Finally, there was a feeling in many circles that saturation was imminent. How long could the UK continue to allow hypermarkets to be built without considerable damage not only to independents and small multiple organisations but increasingly to the large firms themselves?

After the mid-1990s retail environments also became increasingly more competitive in many sectors. In the grocery market we have already mentioned the growth of the discounters entering many Western markets at this time. New legislation across many developed economies in the 1990s and 2000s also allowed a greater market entry into previously restricted markets. Thus, suddenly, grocery firms became bankers and insurance agents; they could sell petrol, open pharmacies and opticians, and sell books and clothing (building on the trend from the 1970s to sell more non-food goods in major superstores).

The distinction between market sectors was becoming increasingly blurred. Greater competition was also becoming apparent through the process of disintermediation, the process whereby manufacturers open their own outlets to cut the profits lost to the retailer or simply keep control of the supply chain process. Benetton is perhaps one of the best known examples (the Italian manufacturer in 2013/4 had over 6,500 franchised stores in 120 countries), but others followed. Nike now has over 300 stores (many of which are factory outlets) including flagship stores in Chicago, New York, Los Angeles and London. Fernie *et al.* (1997) discuss the growth of high brand fashion stores in London, prepared to pay very high rents to get prime retail sites in the West End.

The arrival of tightened planning legislation, greater competition and the threat of retail saturation could possibly have killed new large-scale retail development for good and hence retail location models (mostly applied to out-of-town locations) would consequently become less important. Indeed, a key part of the saturation debate was the belief that it would make store location research less 'sophisticated' in the future. The argument was straightforward – if no new superstores could be built then the more complex methods such as statistical and mathematical models would be less important, replaced perhaps by 'simpler' methods and a major return to gut feeling and intuition. Clarkson *et al.* (1996: 31) for example argued: 'As the UK grocery market becomes increasingly saturated … the need for more sophisticated location assessment procedures become significantly less important'.

However, the reverse has actually proved to be the case – site location teams have had to work harder and more strategically to find ways to grow in these more restricted and competitive markets. And, as Wood and Reynolds (2011, 2012) point out, the sophisticated models are still very much in use. As far as PPG6 was concerned in the UK, in theory, after 1996, it was going to be more difficult to obtain planning permission for major new out-of-town stores. In addition, if the market was really moving towards saturation, it would simply not be profitable to open such large stores in the future. Let's deal with each in turn. First, it should be pointed out that PPG stands for Planning Policy Guidance and hence this is not necessarily the introduction of tough new legislation (see Guy 2006 for an excellent debate on many planning issues in relation to UK retail growth). Local planning authorities were asked to be guided by this legislation, but if a case could be made that there was a need for a new development then this could outweigh the general guidelines of PPG6 (the 'needs test', where evidence could be put to a planning enquiry that an area was suffering from under-provision). Thus store location teams were repeatedly asked by their organisations to put together a case that food retailing in certain locations was inadequate given the size or type of population in a town or part of a city. They could do this by producing many types of provision indicator (store per head of population, floor space per head of population or indeed more sophisticated provision indicators based on gravity style models – see G. Clarke *et al.* 2002 for a review of such indicators, and the discussion in Chapter 5). Thus research into finding good sites for superstores did not stop and a huge volume of planning applications has been lodged since the mid-1990s. [In addition, some companies had sensed that PPG6 would make site acquisition more difficult and, prior to 1996 and the introduction of the new planning guidelines, had bought more land than they could convert into retail space at the time – so-called land banking. Slowly, as the 2000s wore on, they could then convert that land to superstores without further planning consent.]

What about saturation? If the market really was saturated in the mid-1990s then how could any new developments make sense economically? The answer was that the markets were not saturated or anything like it. Until the late 1990s, there was actually little research undertaken to measure the concept of saturation. Langston *et al.* (1997) built on work by Myers (1993) to produce the first maps of provision (per head) across the UK. This work revealed that provision rates were widespread across the UK and it was often in the most densely populated areas where provision was lowest – not in terms of gross floor space but in terms of floor space per head of population. The analysis showed that there were three to four fold differences in provision rates per head from the least to most 'saturated' regions of the UK (see also discussion in Guy 1996, Lord 2000 and an interesting new spin on saturation by Wood and McCarthy 2014). Silcock *et al.* (1999) showed the same widespread variations were present in the US market.

This 'myth of saturation' was also true for non-food retailers. Following investment in more advanced store location research in the 1980s and 1990s, senior managers at the UK book and stationary retailer WHSmith expressed surprise at the opportunities that were revealed through the appraisal of local markets, simply because they had got used to the belief that their network of branches was complete and UK shopping opportunities were saturated. This attitude is common among retailers despite the fact that, in most retail sectors, company market shares vary significantly from region to region and substantially within regions themselves.

The ability of site location teams to address the issues of PPG6 and saturation successfully can be seen in the continuation of the rise and importance of the UK hypermarket and superstore after 2000. Javelin (2015), for example, report the selling space of the grocery hypermarkets went up from 15 to 25 (million sq ft) between 2007 and 2013 with superstores rising from 65 to 79 (million sq ft). In 2008/09 alone for example, Tesco opened 21 new superstores, clear evidence that traditional growth was still occurring. It was not until January 2015 that Tesco announced major store closures for the first time (Telegraph 2015).

Although the continued growth of the superstore allowed retail location teams to continue to search for ideal sites for larger stores, another consequence of the introduction of PPG6 in the UK was for grocery firm's store location teams to rethink the role of the supermarket. The supermarket had been introduced into the UK in the 1940s and 1950s but was overtaken by the superstore from the 1970s onwards. Tesco was at the forefront of such new location research. The store location team now ran their models to test the impacts of opening new supermarkets in many smaller UK towns and cities to see if this format could once again pave a way to grow in the new trading environment. To nullify the impacts of PPG6, these stores were generally below the size limit for superstores (hence around 10,000–15,000 sq ft) and were located in or close to the edge of the high street. Guy (1996) shows how Tesco expanded in Wales using this new supermarket format – the introduction of the so-called 'compact' store. Figure 2.1 shows how this format or fascia was used to spatially infill – that is, it offered the potential to give Tesco a presence in smaller market towns like Aberystwyth, Cardigan and Fishguard, which had previously been felt to have been too small for a major superstore. Thus, rather than a barrier to entry, PPG6 offered Tesco's site location team an opportunity for growth into new spatial markets (undertaking the same analysis in England too).

The battle to convince planners of the need for new developments (the needs test) also began to connect strongly to academic research relating to 'food deserts' – areas of towns and cities where there are low provision rates

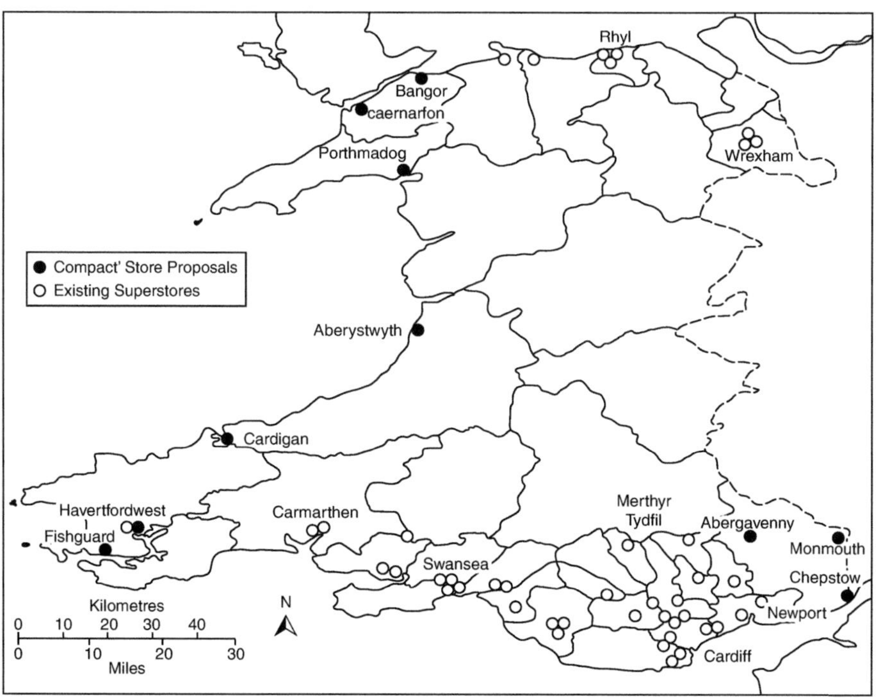

Figure 2.1 The growth of the Tesco Compact store in Wales in the 1990s.
Source: Guy (1996)

of major food stores, making access very difficult for certain consumers, especially those without cars (see Clarke *et al.* 2002; Wrigley 2002; Guy *et al.* 2004; Macintyre *et al.* 2008: see also the use of models to estimate food deserts in Chapter 5). Nowhere is this more evident than in the regeneration agenda now put forward by many grocery firms. If the store location teams can prove there is a need for major investment in certain poorly provided localities in towns or cities, then it may be possible to not only get planning permission for a superstore, but it may be possible to get planning permission for a mega store. The Tesco 'Extra' format, for example, has been introduced in areas of perceived need, many in lower income suburbs largely ignored by the main multiples during the 'store wars' era. These Extra stores are typically 100,000 sq ft or more (a very large store by UK standards). The rationale for such large stores is that the local catchment does not have sufficient spending power (the reason they were ignored in the first place). Thus by building large scale, the retailers can try to generate trade from both within and outside the food desert area. The number of Extra stores has increased dramatically since 1997. In April 2014 there were 247 such stores operational. Interestingly, some of these new retail developments are also linked to new housing or infrastructure to engage more fully with the regeneration process. Other retailers have been keen to join this bandwagon – one of the largest Asda/Walmart superstores in the UK was opened in 2002 in one of the most deprived areas of the UK, the Eastlands region of Manchester, as part of a regeneration package including retail and sport (the Eastlands Commonwealth athletics stadium becoming the home of Manchester City Football Club at the same time). The pressure on store location teams to get these locations right is immense – no one wants a 100,000 sq ft white elephant. (Although interestingly, by January 2015 some retailers were expressing growing concern over the profitability of some of these large stores built in the 2000s.)

2.3 The growth of the discount market

The discount market has traditionally been more important in some countries than others. In the USA for example, firms such as Walmart, Kmart, Target and Costco have been important players since the second world war (see Graff and Ashton 1994; Vance and Scott 1994; Graff 1998, for example, on the rival spatial strategies employed by these competing discount retailers). In Europe, retail growth in the grocery market in the 'store wars, golden era' was fuelled as much by the drive for quality as it was price. Thus, as noted above, a key problem to hit the major grocery players across Western Europe in particular came with the arrival of the deep discount food retailers from Germany and Scandinavia (traditionally stronger discount markets in Europe). In the UK, the discounters located first in areas of major urban deprivation, gaining considerable market share in the north (of England) and the West Midlands. The impact of the deep discounters in the UK was perhaps felt most keenly by the main British incumbent discount retailer who had developed considerable market share in less affluent urban areas – Kwik Save. After its own 'golden period' in the 1980s, Kwik Save's growth slowed down in the late 1990s and it merged with Somerfield in 1996. However, the merger was not sufficient to save the company. On 27 February 2006, Somerfield sold the Kwik Save brand and 171 stores to private equity company 'Back to the Future'. More details on the arrival and diffusion of the deep discounters in the UK can be found in Burt and Sparks (1994, 1995) and Thompson *et al.* (2012).

By the mid-late 2000s there was much optimism in the discount market that future UK growth would be strong in this sector. Between 2007 and 2010, this optimism seemed well placed and indeed was heightened by the onset of recession and the global economic crisis. Similar to the recession in the 1990s, which gave the discounters their initial platform for growth (Burt and Sparks 1994), the more recent recession caused households to switch to the discounters in large numbers. Aldi, Lidl and Netto saw their combined market share rise to 6.1 per cent by 2008, their highest ever in the UK to that date (Aldi 3.0 per cent, Lidl 2.4 per cent and Netto 0.7 per cent : the latter was subsequently purchased by Asda in 2009). Thompson *et al.* (2012) discuss this consumer switch in patronage in more detail, referring in particular to the growth of customers shopping at discounters from the higher income groups. Originally, it was believed that the recession was the sole cause of higher earners reverting to shop at low price retailers. However, the evidence would suggest that the trend was already occurring before 2008 and the recession merely accelerated this trend as households of all types began to seek out low-cost retailers. Aldi opened 50 new sites alone in 2009 and held planning permission on a further 29 sites at the start of 2010 (The Grocer 2010). In 2013 it was the number one UK retailer in terms of percentage sales growth year on year. By 2016 Aldi traded from 660 stores with a market share of 6.0 per cent. Additionally, Lidl held 20 sites with planning permission for new stores at the start of 2014. Thus it seems that growth in the UK market remains very much on the agenda. At the end of 2016 Lidl announced its plan to double its UK store portfolio from around 750 to 1,500 stores. No time for rest in these store location (or marketing) departments! Figure 2.2 shows the location of the UK discount stores in 2010.

Outside the UK it is interesting to see that Aldi in 2013 was also the fourth largest retailer in Germany and Belgium while Lidl was third in Germany and seventh in Belgium, Spain and France.

So what are the implications for store location research? Clearly the discounters themselves worked hard to find ideal sites for their substantial expansion across the UK, and will do so in the future. Examples of the significance of GIS will be given fully in Chapter 3. However, Figure 2.3 shows the usefulness of GIS for overlaying the location of low-income groups with the existing location of discount stores in two UK regions, thus offering site location teams an immediate visual image of possible 'open points' for development.

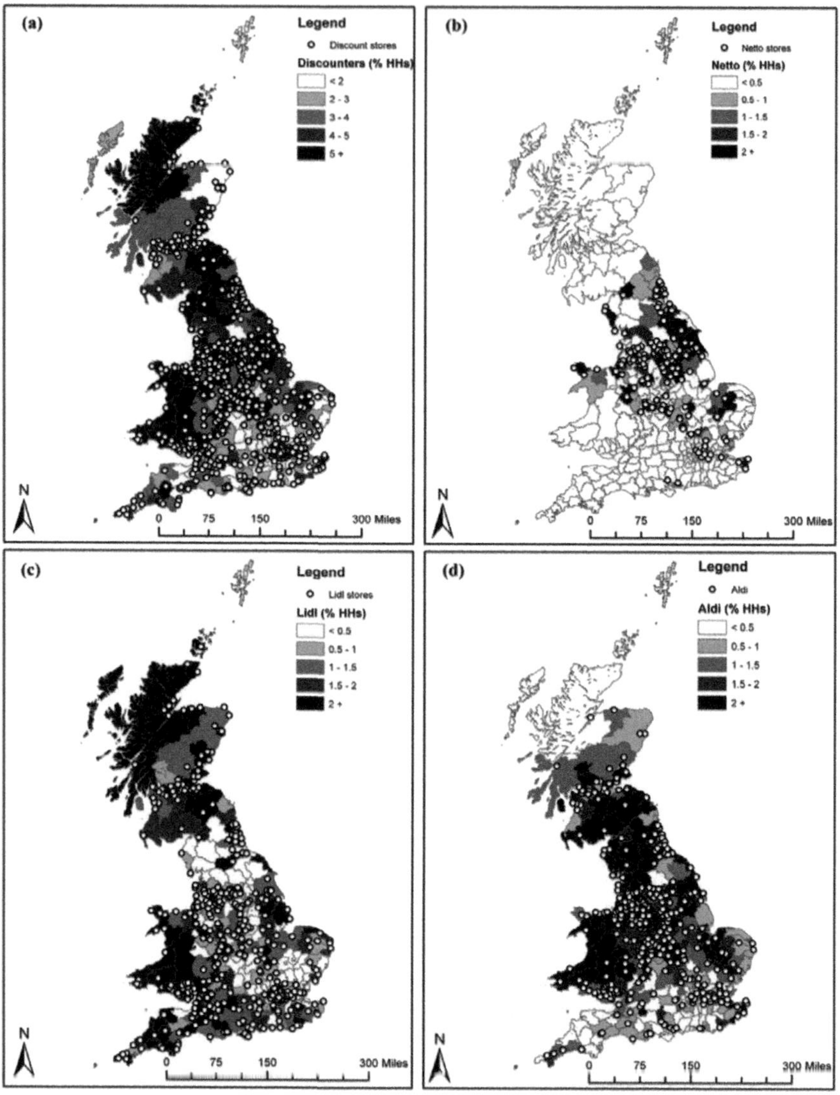

Figure 2.2 The location of UK discounters 2010: (a) all discounters (b) Netto (before sale to Asda) (c) Lidl (d) Aldi.

Source: Thompson *et al.* (2012)

For the mainstream incumbent grocery companies, the growth of the discounters could no longer be ignored. All the main players eventually reacted by discounting prices on main items and there is some evidence of bullying tactics to stop manufacturers supplying the new discounters (and hence a lot

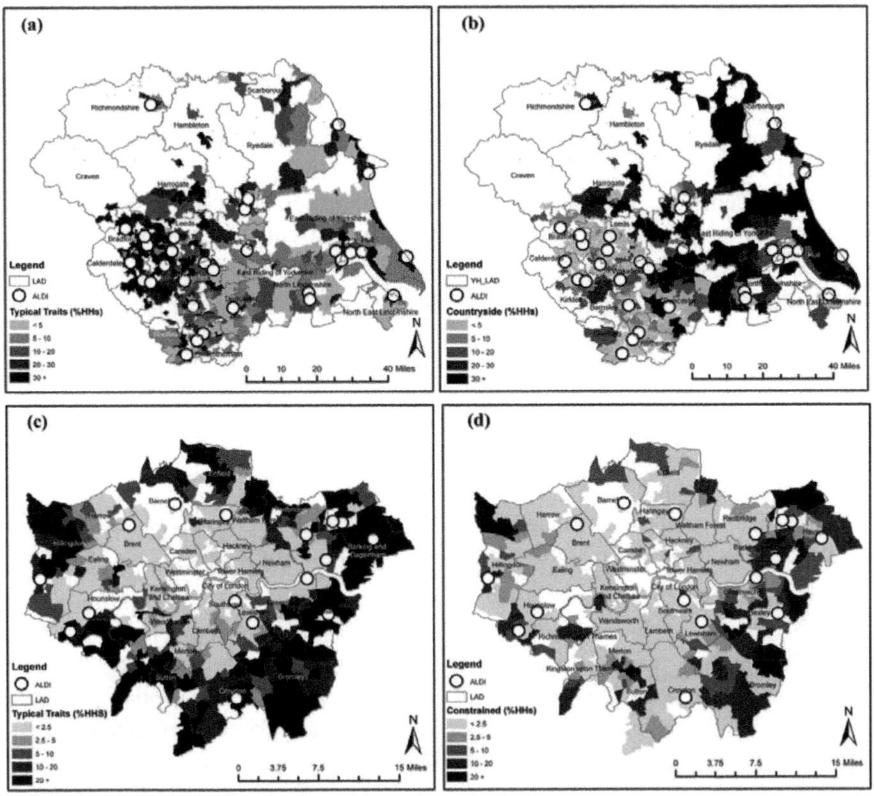

Figure 2.3 Geodemographics by postal sector for Yorkshire and the Humber and London in relation to the location of discount stores: (a) Typical Traits, Yorkshire and Humber (b) Countryside, Yorkshire and Humber (c) Typical Traits, London (d) Constrained Circumstances, London.

Source: Thompson *et al.* (2012).

of products, many unfamiliar brands to British consumers, had to be sourced from Continental Europe). For Asda, Gateway (later Somerfield) and the Co-op, the competition was so fierce in northern towns and cities that they experimented with new fascias (Pioneer for the Co-op, Food Giant for Gateway and Dales Discount for Asda), in order to offer a deep discount format in retaliation. Similarly in France, Carrefour felt the need to create a new format – ED. The ED chain was rolled out across urban France, reaching 897 outlets by 2012. All ED stores were eventually rebranded as Dia stores by Carrefour, who bought the full rights to Dia in France in 2014.

The growth of discount retailing across the developed world has not been confined to grocery retailing. A number of retailers have also exploited the conditions of recession and austerity by growing dramatically since 2000. In the UK we have seen retailers such as Matalan and Primark growing in the clothing market and putting considerable pressure on incumbent high street clothing companies. Matalan have chosen to grow spatially in large shed-like buildings (once seen more typically in DIY-style retailing) in retail parks across the UK. Primark expanded most rapidly in the mid-2000s on UK high streets, especially after the purchase of Littlewoods in 2005 (£409 million). By June 2015 Primark had 167 UK stores and a further 126 overseas.

More recently we have seen a new breed of other 'discount' format shops, some of which have appeared on the UK high street. Following the success of chains of pawnbrokers in the USA, the UK is now seeing the proliferation of shops selling cheap loans, for example the US-owned Money Shop (see Figure 2.4).

These stores are appearing in many UK high streets but are especially prominent in areas of low income, which were more likely to have suffered most in the UK austerity measures introduced after 2010. There is even suggestions that such shops are taking the place of banks – resulting in more informal low-income money markets.

The austerity crisis in the UK is also feeding the growth of another set of discounters that sell a mixture of goods including packaged groceries,

Figure 2.4 New entrants on the UK high street, especially in areas of austerity.
Source: Authors

again typically on the high street. Examples include Poundland, Poundworld, Wilkinson's and Home Bargains. There is even talk of easyJet providing very low-cost food stores: easyfoodstore.com. The first shop was opened in London 2016 selling all items at 25p only.

For these retailers the store location exercise is perhaps simple – to find the communities around the UK that are suffering most from high unemployment and declining incomes (either from reduced benefits or pay freezes in the public sector) as the recession bites hard. It would also be an interesting research question to explore the correlation between the austerity measures, bank branch closures and the rise of new money-lending stores.

2.4 The convenience market and opportunity for new store locations

In Chapter 6 we shall discuss the growing demand for convenience in retailing as household types change and more people are becoming time conscious. This change in demand coincided with the tougher retail operating environment of the 1990s and 2000s. Some of the large food retailers in particular, however, saw the opportunity to combine these issues and from the mid-1990s onwards they began to research the convenience market (generally defined as stores of less than 3,000 sq ft – a size threshold that allows them to escape Sunday trading laws and open longer hours; although as we write in 2015 this is under review in the UK). In fact, so important was this sector to growth after 2000 that many store location departments split their activities– to deploy dedicated teams to explore the geography of the convenience market. The convenience market in the UK is estimated to take about 20 per cent of the total grocery spend (IGD 2015), although interestingly this had previously been treated as a very separate market compared to the main superstore market (most notably by the UK Competition Commission who in 2000 declared them to be separate markets, thereby giving a legislative green light to the major players to expand their convenience store networks).

The UK grocery convenience market was reported to be worth £37.7 billion in 2015, having grown from £19 billion in 2000. It is expected to reach $44 billion in 2017 (IGD 2015). Yet, to date, little has been written on the changing nature of the convenience market and how the recent entrance of the largest UK grocery firms has produced a fascinating spatial battleground across the UK. In this section we follow the argument and discussion presented in Hood *et al.* (2016). The main providers of convenience retailing during the golden era of superstore development were mainly the independents (see also discussion on symbol groups below). Baron *et al.* (2001: 398) discuss the traditional advantages of independent retailers over their large corporate rivals: 'their convenience in location and opening hours, home delivery, friendly and personal service and informal financial services such as extended credit and Christmas clubs'. Post-1996, however, there is little doubt that the independent sector has generally suffered, not only due directly to the continuing development of superstores but also to the rise of new players in the convenience market itself. This has been especially notable in the decline of independent butchers, bakers, fishmongers, fruit and veg shops etc., especially in traditional high streets and parades (though see Wrigley *et al.* 2009; Wrigley and Dolega 2011; Lambiri *et al.* 2016 for a more nuanced discussion of the potential impact of the major grocer firms on UK high street convenience retailing).

Despite this decline, the independent sector still had 33 per cent of the convenience market in 2000 (Mintel 2003). Yet even that considerable market share was to be eroded away post-2000.

Increasingly, it has been recognised that independents were facing growing problems around the complexities and inconsistencies found in the supply chain, low operating margins, lack of sufficient capital for investment, difficulties around the administration of value added tax and a lack of business experience (problems raised by Dawson and Kirby 1979, discussed again in Baron *et al.* 2001). By 2012 the market share of the independents in the convenience market had slumped to 19.4 per cent (IGD 2015). However, part of that decline is misleading as a number of independents did in fact join the so-called symbol groups. Symbol groups can be defined as umbrella retail organisations under which a number of convenience stores operate taking advantage of a branded shop fascia, advantageous buying terms (the scale economy effect of the symbol group purchasing in bulk), access to own brand ranges, IT, logistical and marketing support, and general advice and guidance. In 2000, 6,900 convenience stores were operating under symbol group branding in the UK, rising to over 17,000 by 2014.

As noted in Section 2.1, faced with tightened planning legislation in the mid to late 1990s, the major superstore players adopted many different strategies for growth including the development of convenience stores (Wood *et al.* 2006). In addition to overcoming the new tightened planning regulation, convenience stores can offer attractive margins and sites have become relatively easy to obtain as a number of traditional UK high street retailers have gone bankrupt and sold sites to the multiple grocers (including many pubs in the UK, which have closed and offered grocery retailers sites of a considerable size with existing planning consent to sell food – see Smithers 2012). Table 2.1 summarises the involvement of the major players in the convenience market in 2012.

Figure 2.5 shows the spatial extent of the UK grocery convenience market in 2012. Not surprisingly, when plotted as raw numbers, the highest floor space totals (>650,000 sq ft) are in the key urban areas. Notably, Glasgow (G), Newcastle upon Tyne (NE), Sheffield (S), Nottingham (NG), Birmingham (B), Swansea (SA) and Cardiff (CF).

Figure 2.6 plots the spatial variations in the market share of the convenience market, expressed as a percentage of all floor space in the grocery sector (cf. Hood *et al.* 2016). When expressed in this way a very different pattern emerges – high market shares (often with low total floor space) can be seen

Table 2.1 Major convenience retailer store numbers and market share, 2012.

Retailer	*Store numbers*	*Market share (%)*
Co-op	2,170	14.1
Tesco	1,946	10.0
Sainsbury's	425	2.0
Musgrave Group	2,006	13.8
Costcutter	1,546	10.3
Premier	2,670	19.2
SPAR	2,232	14.0

Source: Hood *et al.* (2016)

in the more rural areas. In the more rural retail landscape in north and west Wales for example, branded fascia convenience stores featured prominently and accounted for over 30 per cent of total grocery floor space in 2012. Conversely, branded convenience grocery stores had a market share of less than 15 per cent in a number of postal areas in 2012. These include the more urbanised postal areas such as Sutton (SM) and Twickenham (TW) (both in London), Halifax (HX), Liverpool (L) and St Albans (AL). These are areas that the major convenience retailers might exploit more in the future as they continue to look for spatial expansion.

As shown in Table 2.1, a leading major convenience retailer is the Co-operative Group, a consortium of 22 different societies across the UK (although each has its own name we shall look at the combined market share under the banner of the Co-op). Figure 2.7 shows the market share of the Co-op for both the convenience market and the total grocery market. The Co-op has historically made the greatest commitment to growth through small-format retailing. In 2002, the Co-op acquired the Alldays brand of 600 convenience stores, becoming the largest convenience retailer among the major grocery firms in the UK, with over 2,200 convenience stores. In 2003, the Co-op's convenience arm continued to growth through the acquisition of Balfour, a convenience store chain with 121 stores. However, in 2012, the Co-op acquired 880 stores from Somerfield expanding its offer in the small to medium grocery store offer. It could be argued that this signalled the retailer moving away from its earlier primary commitment to the small-store convenience market alone.

Figure 2.7(a) shows that the Co-op is well represented across the UK for both the total market and the convenience market. Given that many of Co-op stores are less than 3,000 sq ft, it is not surprising that the two maps are similar in terms of spatial patterns. It is interesting that the impact of the Somerfield purchase is evident in Figure 2.7(b) as Somerfield had traditionally being strong in Wales and the west of England. The Co-op comprises a consortium of different companies and the most powerful of these is the Co-operative Group (which merged with the second biggest Co-op 'United Co-op'

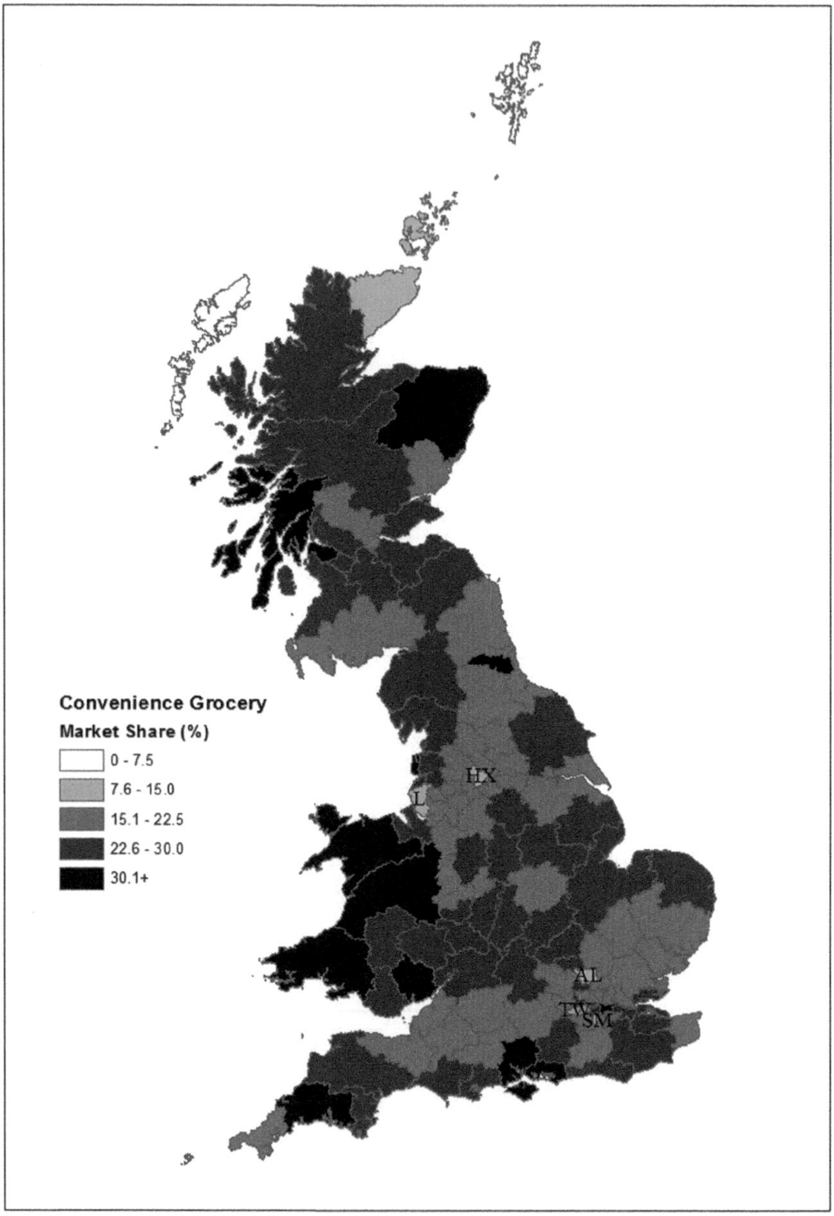

Figure 2.5 Convenience grocery floor space by postal area in Great Britain, 2012.
Source: Hood *et al.* (2016)

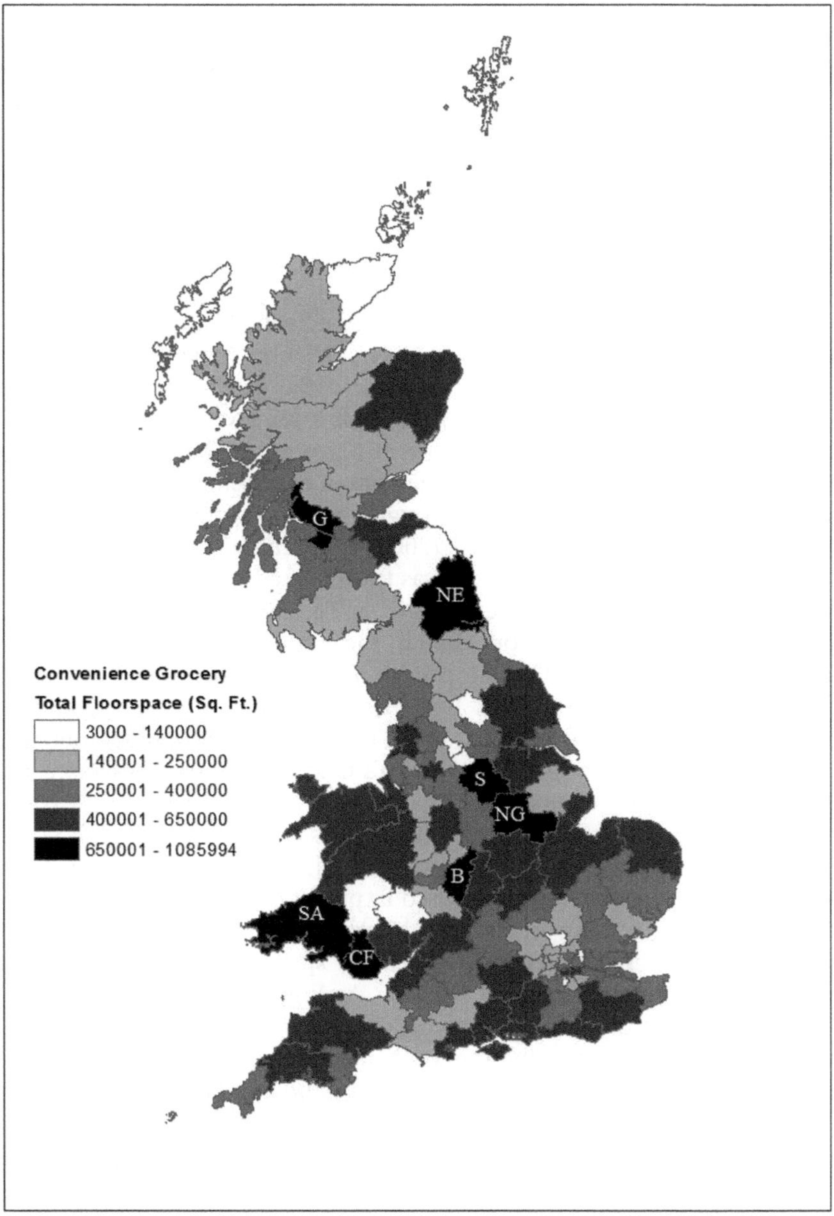

Figure 2.6 Convenience grocery floor space share of the total grocery market, 2012. Source: Hood *et al.* (2016)

in 2007), the East of England, the Midlands, Southern and Scotmid. This can be seen in the pattern of high market share seen in Figure 2.7(b), especially in Lancashire and Yorkshire (the Co-operative Group), Essex and Suffolk (East of England), Staffordshire and the Potteries (Midlands), Scotland (Scotmid). As seen in Figure 2.7(a), in 2012 the Co-op also had a large share of the convenience market in rural postal areas in northern Scotland. Moreover, the retailer had a large convenience market share in much of northern England, including North and West Yorkshire, Lancashire, north-west England including Greater Manchester and large parts of the south coast including Brighton (BN), Portsmouth (PO) and Southampton (SO), showing the influence of Southern Co-op (as seen in Figure 2.7(a)).

As shown in Table 2.1, Tesco is also at the forefront of convenience retailing. Figure 2.8(a&b) shows the market shares of Tesco for both the convenience market and for the total grocery offer. In 1994, Tesco undertook its first foray into convenience store retailing through a joint venture with Esso to open, branded convenience stores on petrol forecourts. This proved successful and the retailer continued to pursue convenience retailing through both forecourt and non-forecourt stores through a new format and fascia – Tesco Express (see Figure 2.9). Wood *et al.* (2006) argue that the competitive landscape of the convenience store sector was transformed in January 2003, when Tesco purchased 862 convenience stores from T&S Stores, boosting the total number of small-format stores operated by the retailer to around 1,000. These stores retained the original One Stop store branding under which they were previously trading. Additionally, Tesco acquired the London-based convenience store chains Europa, Harts and Cullens from their parent company Administered in 2002. In late 2010, Tesco's One Stop brand purchased the Mills chain of 76 convenience stores operating in the Midlands, south Wales and the north-east of England, increasing Tesco's One Stop chain to 598 convenience stores. By April 2015, Tesco had a total of 2,500 convenience stores when combining the One Stop and Tesco Express fascias.

Tesco, with 28 per cent of the total grocery market in 2016, is the most national of all UK grocery retailers in terms of spatial coverage. Tesco's convenience stores are more clustered spatially. As Figure 2.8a shows, Tesco has its largest share of the market in the postal districts in the south of England, particularly around London and the south-east. Conversely, Tesco has a relatively low market share in Wales, northern Scotland and north-east England.

We have seen that a true battle has emerged between retailers vying for large market shares in convenience grocery retailing, in which geography plays a key role. As more retailers widen their portfolio of convenience stores, competition is likely to intensify, placing increased pressure on both existing stores and on maximising the effectiveness of location decisions for new convenience stores. For example, the other main UK grocery retailers are now looking to the convenience market for expansion. Waitrose opened its first small-format store in 2011, branded Little Waitrose, in South Kensington (Whiteaker 2011). After local trails in various locations, Asda have recently announced

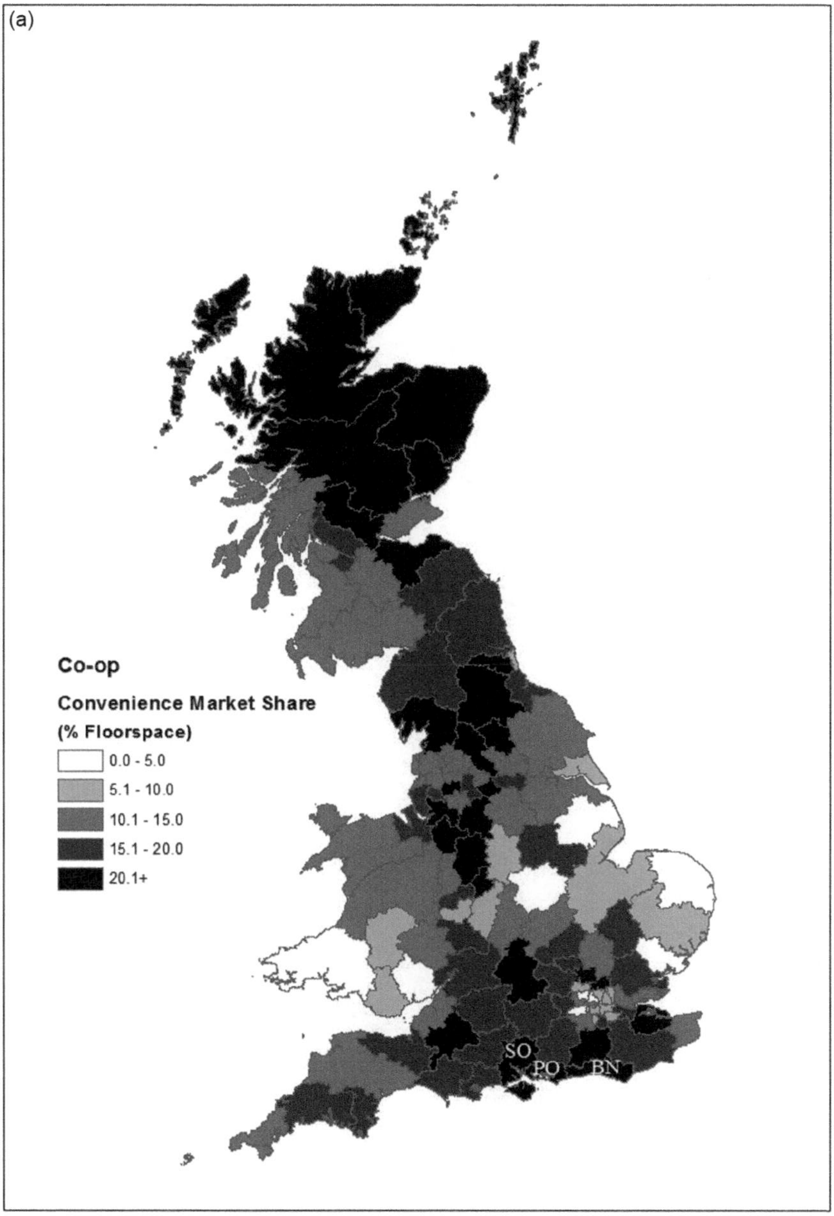

Figure 2.7 Market share of the Co-operative Group by postal area, 2012: (a) convenience (b) grocery.

Source: Hood *et al.* (2016)

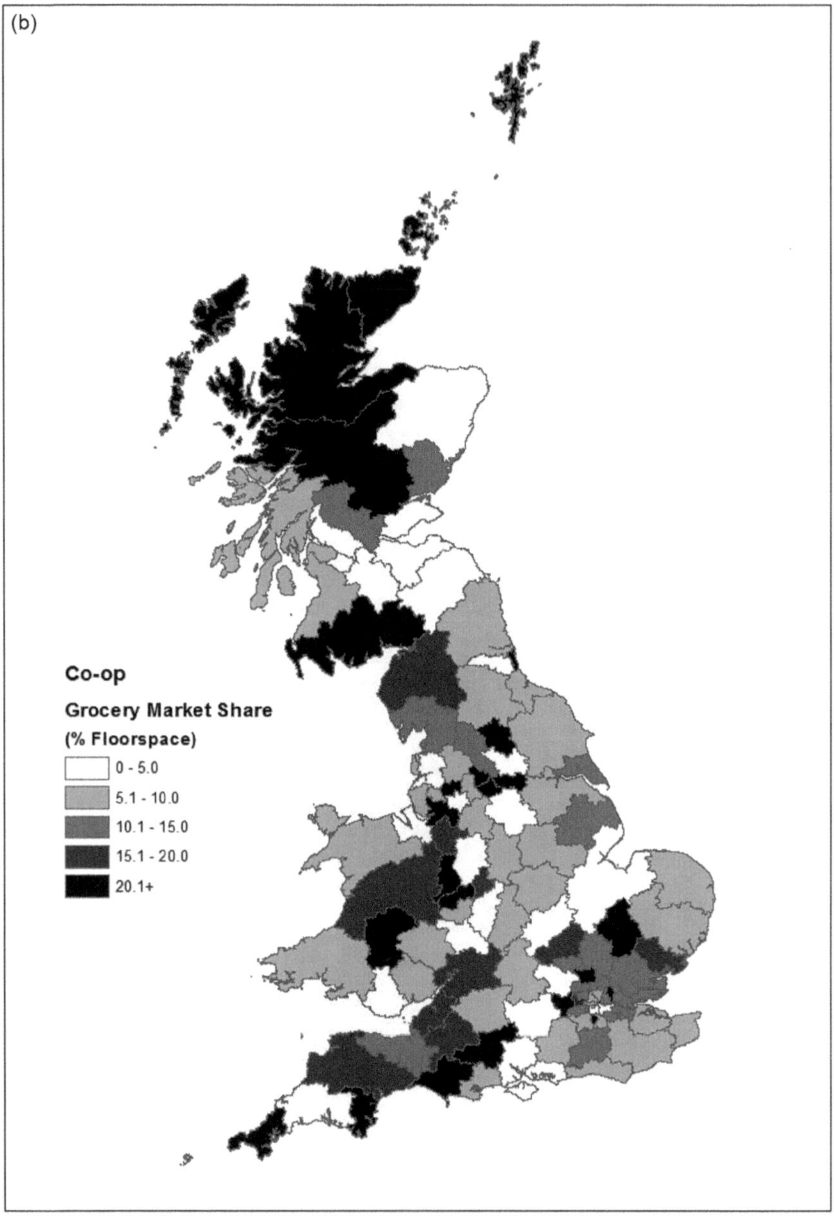

Figure 2.7 (cont.)

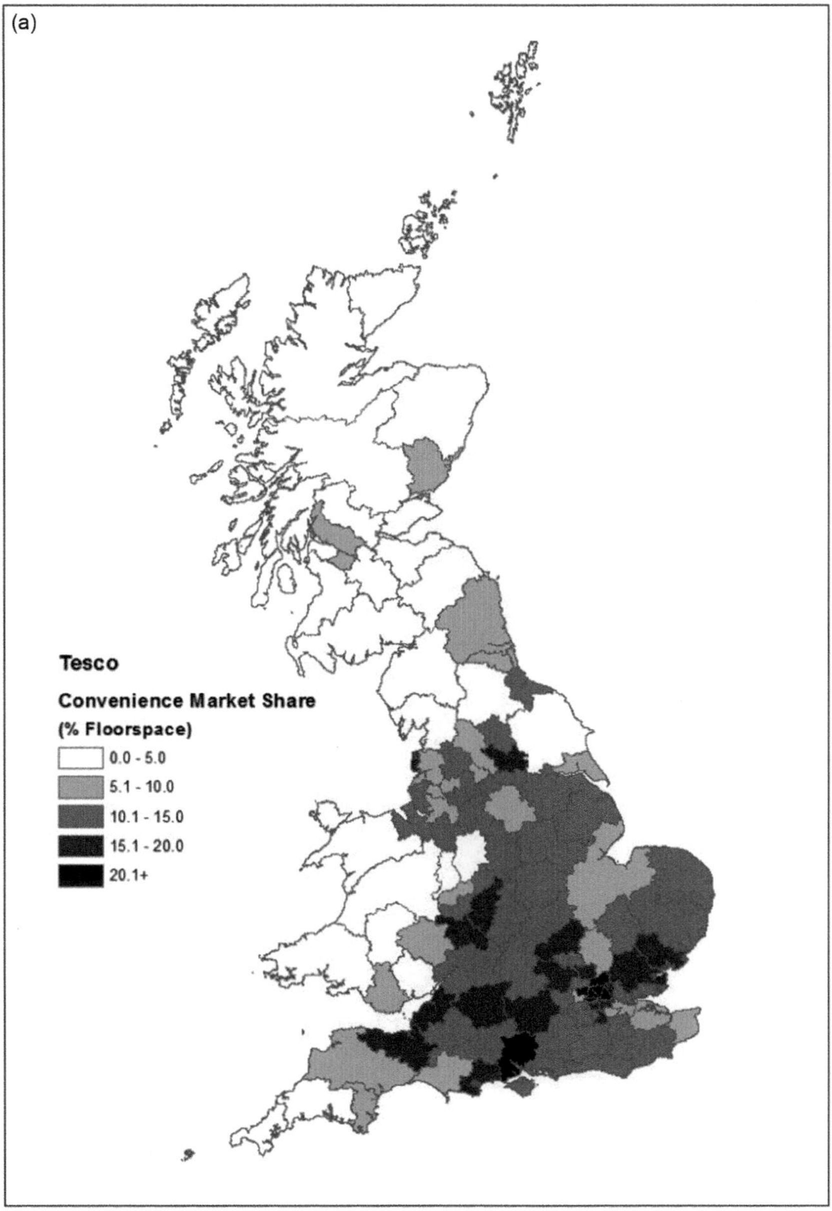

Figure 2.8 Market Share of Tesco by postal area, 2012: (a) convenience (b) grocery.
Source: Hood *et al*. (2016)

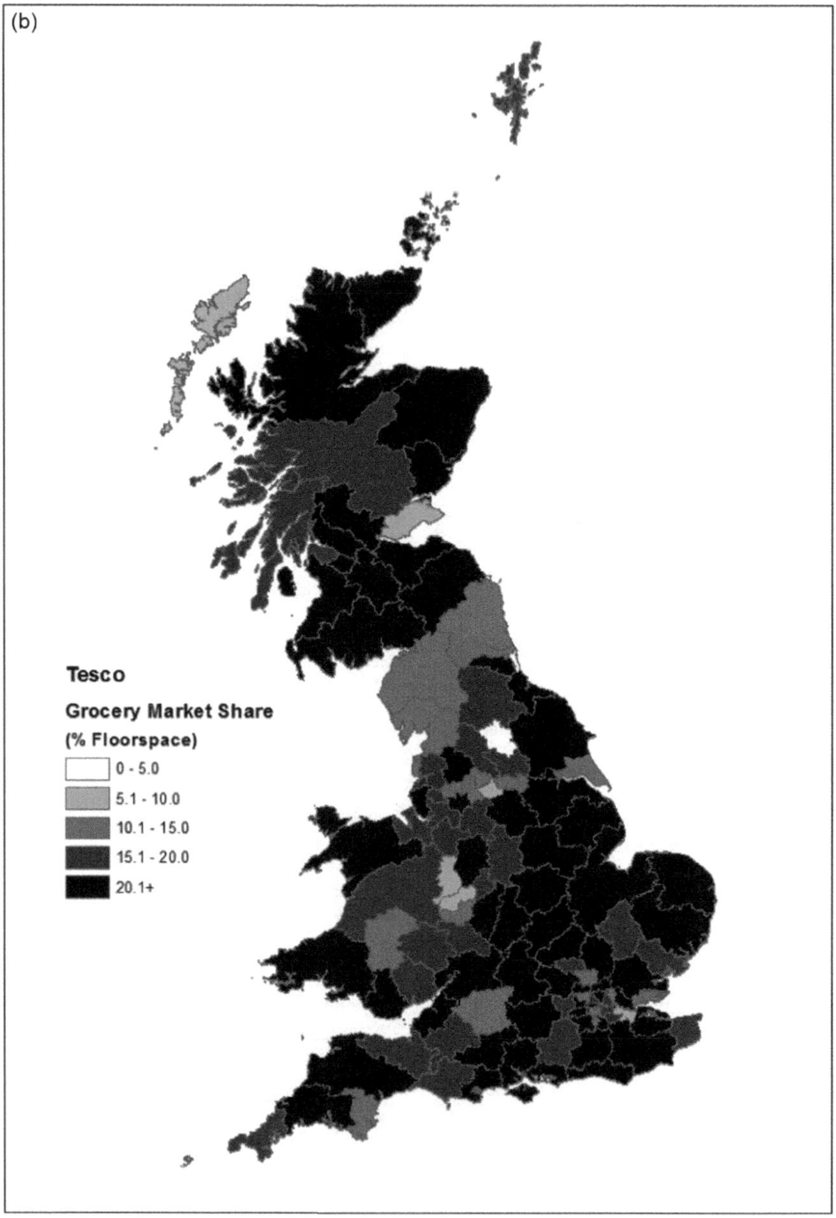

Figure 2.8 (cont.)

Figure 2.9 Tesco Express: their main format for convenience retailing.
Source: Authors

their preferred strategy of growing a c-store format through existing (and new petrol) station sites (combining this with their 'click and collect' e-commerce offering) and, in the London area, through the network of underground stations (see also discussion in Chapter 6 and Chapter 10 regarding rail stations and their potential). Although also a relative latecomer, Morrisons committed to developing their offer in the small food-retail-based convenience store sector in 2012. Dalton Philips, Chief Executive of Morrisons announced, 'Convenience is one of the fastest growing sectors of the market and developing our offer in this channel is a key part of our growth strategy' (Morrisons Online 2011). The retailer opened its first two convenience stores in 2012, branded M Local, and held talks with Costcutter with the intention of purchasing a large number of small-format stores in 2014 to advance their convenience store offer (Leyland *et al.* 2012). In early 2013, Morrisons also acquired a number of retail units previously operated by Blockbuster, Jessops and HMV (all non-food retailers which have recently failed on the high street) confirming the retailer's (then) intentions to advance their convenience offer (Neville 2013). However, concerns are already being expressed about market saturation in the convenience market. Generally poor trading patterns in 2014/15 made Morrison's decide to sell off its new convenience stores, and to refocus on its core superstore business.

To add to what will be a very exciting future battleground across space, we should not forget the arrival of a new breed of overseas convenience retailers to the UK – ethnic retailers from East Europe and the Middle East (Guy 2008; Wrigley *et al.* 2009; Wrigley *et al.* 2017). Although small in total, these are becoming important players in some regional markets of the UK. This new battle for the convenience market is also mirrored in other developed world countries. Carrefour's Express stores now total 300 plus all over France.

What are the implications of these new formats for site location research? According to Wood and Browne (2006, 2007) and their interviews with site location analysts, the traditional techniques of market analysis for large-scale food stores will become largely redundant, replaced with a 'back to basics' approach to market analysis; that is, the use of more site visits in combination with simpler quantitative techniques. Thus the perception in the industry is that the gravity or spatial interaction model and other modelling techniques are less useful for convenience locations, and many firms use simpler forms of GIS (see Chapter 3 for more discussion). Often, the store location team try to use convenience stores to spatially infill around their main superstores. Whether or not spatial models can be effective in the convenience market has not been researched to date, or at least such findings have not been published. This remains an ongoing research question but preliminary findings can be found in Hood *et al.* (2017).

The advancement of Sainsbury's into the convenience market was accompanied by a partnership with Shell UK. The alliance resulted in the retailer opening 100 convenience stores on petrol forecourts across the UK, although this partnership has subsequently disbanded. This introduces another task for the store location teams – how to gain an additional market share via the petrol forecourt market. The latter was worth £4.1 billion in 2014 (IGD 2015) and there were about 9,000 petrol stations operating in the UK. It is interesting to speculate on ideal partnerships and what-if scenarios involving the food retailers and the petrol companies. Figure 2.10a shows the estimated market share for groceries of Sainsbury's in 2010. The dominance of its share in the south-east and London is notable (the darker shading represents approximately 30 per cent of the UK grocery market). If Sainsbury's had progressed the partnership with Shell, and opened a convenience store in every one of Shell's petrol stations then Figure 2.10b shows the new market share that would be possible (this is based on the outputs of a spatial interaction

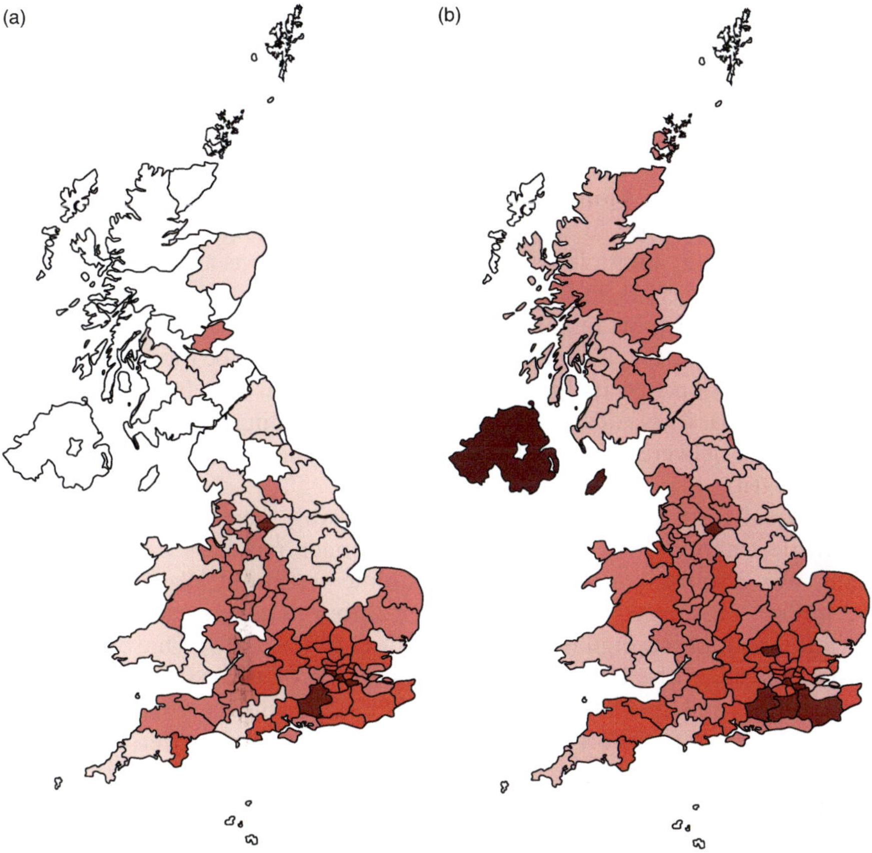

Figure 2.10 Sainsbury's market share (a) without trading at Shell stations (b) with trading at all Shell stations (dark shading = 30%).

Source: Authors

model run for the entire UK grocery market). Given Shell's presence in the north of England, Scotland and Ireland the spatial fit would have been a real bonus to Sainsbury's aim of becoming a truly national UK retailer.

2.5 Exploiting new locations

In this section we explore the way in which some retailers are exploiting new types of geographical spaces to trade given the more competitive environment described in Section 2.2. First we look at the growing interest in transport hubs. Then we look at tourist sites and finally we consider site location in international markets.

2.5.1 Transport hubs

The desire to win new customers has seen retailers trying to exploit new spaces which so far have been relatively ‘untapped’. The most obvious case is the growth in airport retailing. Eurostat (2015) shows the number of passengers in 2013 recorded at principal airports across West Europe. Given these size of these numbers (i.e. London Heathrow 72 million, Amsterdam Schiphol 52 million, Munich 37 million), and the fact that potential shoppers are effectively trapped for a number of hours in waiting areas, it is not surprising that many multiple retailers are seeking locations at airports. Freathy and O’Connell (1998) and Thompson (2007) provide a good discussion and introduction to airport retailing.

In addition, a few airports are now being used as locations for ‘virtual stores’. Tesco opened its first at Gatwick Airport in London in 2012. The idea is that passengers can place an order via a vending-type machine which will then be delivered to their home address on return from their holiday or business trip.

After airports, railway stations are becoming increasingly popular as retail destinations. As with airports, passenger numbers can be significant. For example, in the UK, the mainline station at Waterloo receives 62 million passengers each year, while Edinburgh enjoys a throughput of 23 million and Birmingham around 8 million (ORR 2016). One of the main grocery firms in the UK to exploit the potential of the railway station has been Marks and Spencer – keen to roll out its convenience format by tempting commuters with lots of ready meals and wines etc. after a busy day at work. Marks and Spencer entered the convenience sector through its ‘Simply Food’ fascia, with its first in Twickenham, London in 2001. The retailer has continued to grow its operations through stores on major transport links and in 2013–14 operated 420 Simply Food convenience grocery stores nationally, with 38 located on train station concourses. In 2015 it also had 120 located at BP petrol stations.

In similar fashion retailers are increasingly appreciating the potential of the London underground stations as prime retail space. The annual passenger numbers are again extraordinary: for example, Liverpool Street 63.65 million; London Bridge 65.44 million; Oxford Circus 77.09 million; King’s Cross/St. Pancras 77.1 million; Victoria 82.25 million and Waterloo 84.12 million (Transport for London produces annual estimates of station usage on their website – TFL 2016). Sainsbury’s have targeted a number of their ‘Central’ convenience stores as close to underground stations as possible, not only to tap into tube travellers but to target the large numbers of shop and office workers in the catchment areas of the tube stations. Carrefour too has a new ‘City’ format to capture workers in major urban areas.

Finally, even motorway service stations are becoming attractive locations for retail development. Waitrose has launched a joint venture with Welcome Break to open convenience stores in their service stations while Moto and

Marks and Spencer announced a similar £590 million partnership deal in 2010. By 2015 there were 38 'Simply food' stores at Moto service stations.

2.5.2 Holiday resorts

The growing importance of tourist demand in certain regions, and the need for retailers to start taking this more seriously in their store location research, has also seen a drive to new locations. Most tourist regions have a varied stock of visitor accommodation, encompassing a range of locations, accommodation types, grade/rating and price. If camping/caravanning and holiday parks play the dominant role in the overall provision of tourist accommodation in a region, such accommodation can be expected to generate a high grocery (and related products) spend provision as such accommodation is geared heavily towards self-catering. A key question is how much additional spend do they bring to local shops and businesses, especially in the summer period?

Fortunately there are a number of tourist surveys which help answer this question. Newing (2013) demonstrates the amount of expenditure associated with different forms of holiday accommodation in Cornwall, UK. Overall, average total trip expenditure is highest among visitors using self-catering accommodation, which is unsurprising given the higher length of stay among visitors using this form of accommodation. Retailers are therefore looking to add tourism estimates into existing store location revenue estimations (Newing *et al.* 2013a, 2013b, 2014 show one method for achieving this). Some have gone one step further – negotiating deals for exclusive rights to operate convenience stores within holiday parks. In 2010 for example, Spar and Bourne Leisure agreed a deal to allow Spar exclusive rights to provide convenience stores in 46 Butlins and Haven holiday parks owned by Bourne.

2.5.3 International growth

Another major store expansion strategy adopted by many retail firms under the conditions of harsher planning regimes, increased competition and saturation at home has been to expand internationally. Carrefour chose to expand through a policy of 'contagious diffusion', locating first in neighbouring countries such as Belgium, Spain and Italy, before then expanding significantly in Latin America and Asia. Tesco chose the more liberated planning regimes of East Europe to make its first moves abroad. The literature on retail internationalisation is now vast. There are many papers on why retailers have chosen to expand internationally and the ways they have achieved this (i.e. Choi *et al.* 2002; Wrigley *et al.* 2005; Dawson *et al.* 2006; Burt *et al.* 2008; Alexander and Docherty 2009). There are also a wealth of very useful case studies (Goldman 2001; Fernie and Arnold 2002; Palmer 2005; Coe and Lee 2006, 2013). More recently, there has been a number of studies focusing on problems and failure in international markets (Burt *et al.* 2003; Wrigley and Currah 2003; Christopherson 2007).

There is a paucity of literature, however, on how firms adapt their store location techniques when researching overseas markets. The common belief is that in many markets, especially in the developing world, data is hard to come by, and tends to be incomplete both temporally and spatially. If censuses exist, for example, they tend to be incomplete especially in more rural and impoverished regions. Thus, many location teams argue they are forced to return to very basic methods of store location – perhaps simple GIS functions related to buffer and overlay (see Chapter 3). However, Meksangsouy (2012), for example, showed that it was possible to build the necessary data sets from a variety of public and private sources in Thailand and was able to produce extremely well-fitting spatial interaction models for the Bangkok region (see also the discussion of international geodemographic systems in Chapter 4). Figure 2.11 shows the market shares predicted by the models for different retailers in Bangkok based on these models (we shall discuss the workings of these models in Chapter 5).

2.6 Conclusions

This chapter has discussed the changing trends in retailing that have had a profound impact on the business of store location research. Despite a belief that site location might be simpler in an era of more restricted, saturated markets the opposite has proven to be true. The site location teams of many of the major retailers have never been busier – and indeed many have expanded considerably in terms of numbers of analysts employed. Since the tighter planning regimes of the late 1990s, retailers have found new ways to grow, alongside maintaining, to some degree, traditional growth models. This growth in UK grocery stores since 2001 has involved a mixture of traditional superstores, discount supermarkets and convenience outlets. Thus, despite the slowing down of superstore developments in the UK in the 2010s this chapter has shown plenty of evidence of how busy site location teams remain.

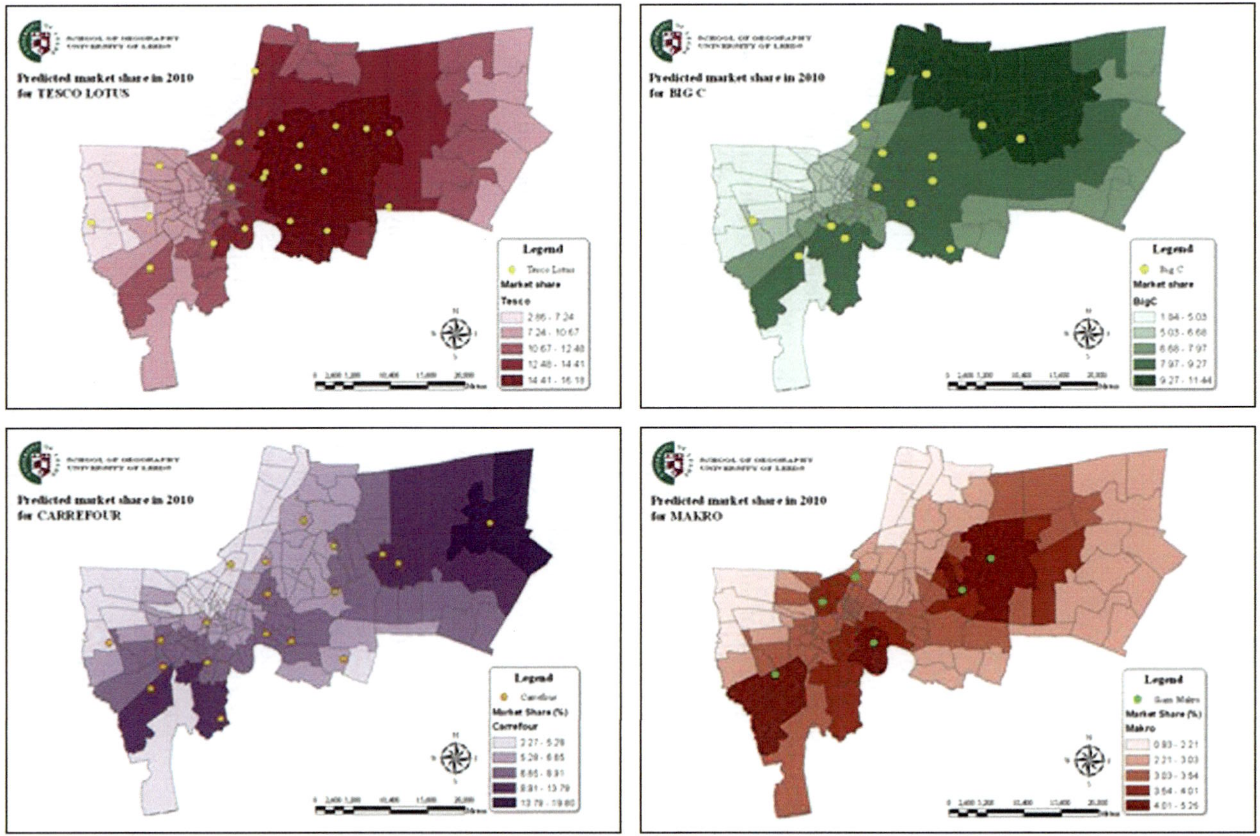

Figure 2.11 Estimating market shares for the then incumbent grocers in Bangkok in 2010.

Source: Meksangsouy (2012). Reproduced with permission from Pakorn Meksangsouy

3 GIS and models for retail planning and analysis

3.1 Introduction

In Chapter 2 we looked at the locations providing new opportunities for retail growth and the implications of this for store location research, especially from a UK perspective. The principal search for new out-of-town locations common in the 1980s and 1990s has been replaced, or at least supplemented by, the search for new types of location to serve more niche markets. As retailers have increased the number of store formats on offer so the store location research exercise has arguably become more diverse and interesting from a spatial perspective. In the next few chapters we explore the common methodologies used in store location research, updating the frameworks and examples introduced in Birkin, Clarke and Clarke (2002). In this chapter we first explore the use of geographical information systems or GIS for retail planning and location analysis. GIS has become widely established in many areas of public and private sector planning over the last 20 years or so. A GIS can be defined as: 'a system for capturing, storing, checking, integrating, manipulating, analysing and displaying data which are spatially referenced to the Earth' (Department of the Environment 1987: 132)

Useful reviews of GIS in human and physical geography appear in a number of excellent introductory texts (i.e. Longley *et al.* 2005; Heywood *et al.* 2011; Ballas *et al.* 2017). A GIS typically stores data relating to many different attributes. This might include data on geographic location (a postcode, census tract or Cartesian co-ordinate for example) and variables relating to features at that location: a grocery store of a certain brand and size for example. Given a number of data layers, a user can interrogate the data within a GIS to find out the characteristics of features at a particular point, or conversely, find all locations that satisfy certain pre-set search criteria; that is, find the locations of all Kroger stores over 50,000 sq ft. It is this ability to link data sets which makes GIS so useful. Retailers have also been keen to use GIS to aid the store location decision-making process and to help with local marketing strategies. This chapter also builds on, and updates, the discussion of Benoit and Clarke (1997) and their appraisal of the usefulness of GIS in store location research. In Section 3.2 we explore the use of GIS for geocoding and mapping. Buffer and overlay analysis is discussed in Section 3.3, especially in relation to the problem of how to undertake store revenue forecasts using this technique. In Section 3.4 we introduce network analysis within a GIS framework and how this has been able to aid the understanding of both access to retail facilities and help estimate real distances travelled – a growing research agenda for estimating food miles and carbon footprints within retail distribution.

3.2 GIS for mapping

Arguably, the greatest attribute of a GIS is its ability to first store and then plot geographic information on a map. Maps can bring data to life, especially when geographical information is otherwise lost or hidden within spreadsheets or databases. Table 3.1 for example shows the market share for the grocery retailer Asda estimated for a selection of postal district areas in Leeds in the UK. Thus, Asda is estimated to have 65.88 per cent of the money spent by residents in LS14 (the location of a major Asda/Walmart supercentre in east Leeds). This information is obviously useful in that postal districts of high and low market share can be immediately seen. However, the spatial patterns in the data are not evident from the table as it is not possible to see which postal districts are neighbours of others. For example, LS14 and LS15 are in the east of Leeds while LS16–20 are located in the north-west and high market share is driven by proximity to a different Asda store. Figure 3.1, on the other hand, maps this data and shows the spatial patterns of high and low market share for Asda – this time for the whole of Leeds and Harrogate. The location of stores (*) can also be plotted so that the relationship between market share and individual store location can be observed and understood better.

As well as market share data, GIS can help map census data which helps identify variations in retail demand. We shall discuss the estimation of demand more fully in Chapter 6. Plotting census variables is common practice in retailing. Figure 3.2 (adapted from Benoit and Clarke 1997) shows the distribution of the lowest income groups in Leeds, as represented by social class groups D and E (unskilled, manual workers), alongside the location (.) of the discount retailer Kwik Save (before its demise in 2008) who were keen to

Table 3.1 Estimated market share for Asda in postal districts around its east Leeds store.

Asda market share	
LS14	65.88
LS15	26.08
LS16	35.78
LS17	14.26
LS18	14.31

Source: Authors

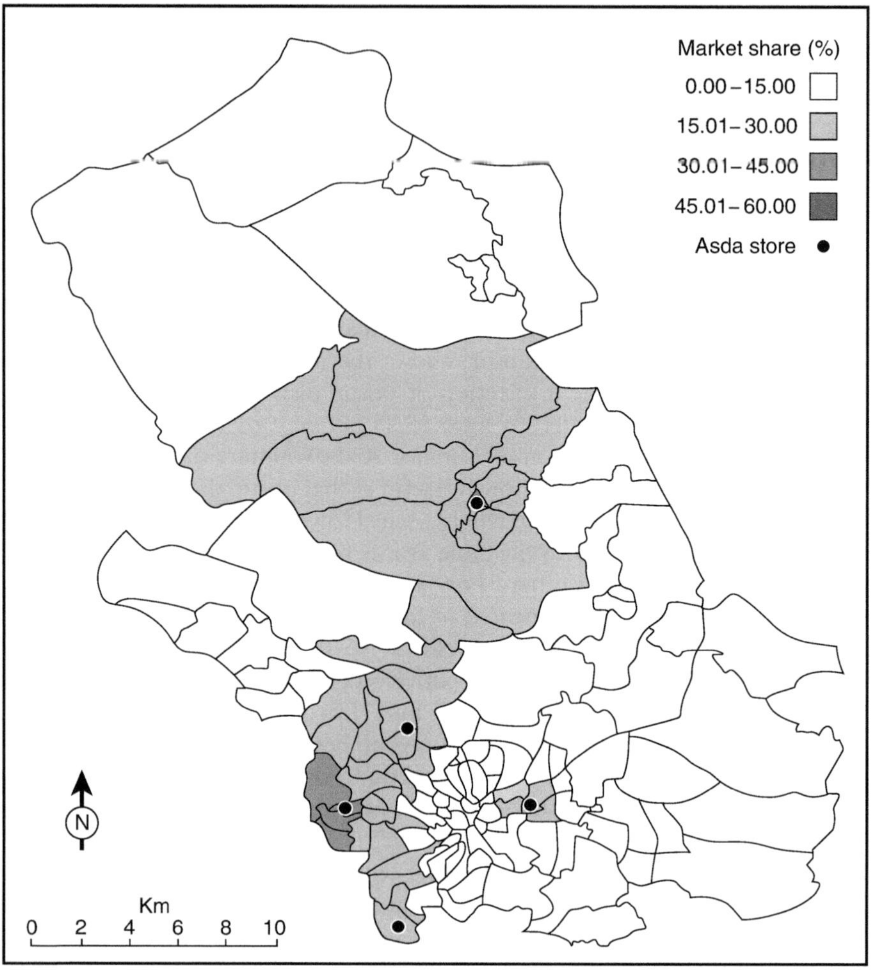

Figure 3.1 Asda's estimated market share across the Leeds and Harrogate areas.
Source: Authors

target such demographic groups. Thus for example, it is easy now to visualise the location of the population most likely to be customers of discount retailers. By overlaying the existing discount stores it is possible to see 'open points' and possible good locations for new stores.

Maps of age or income etc. are often plotted alongside store locations and perhaps major infrastructure such as roads and bus routes. The availability of Google Maps (and a whole host of similar products) now allows the routine mapping of detailed road networks alongside retail attribute data. The use of static two-dimensional maps in GIS has been increasingly supplemented

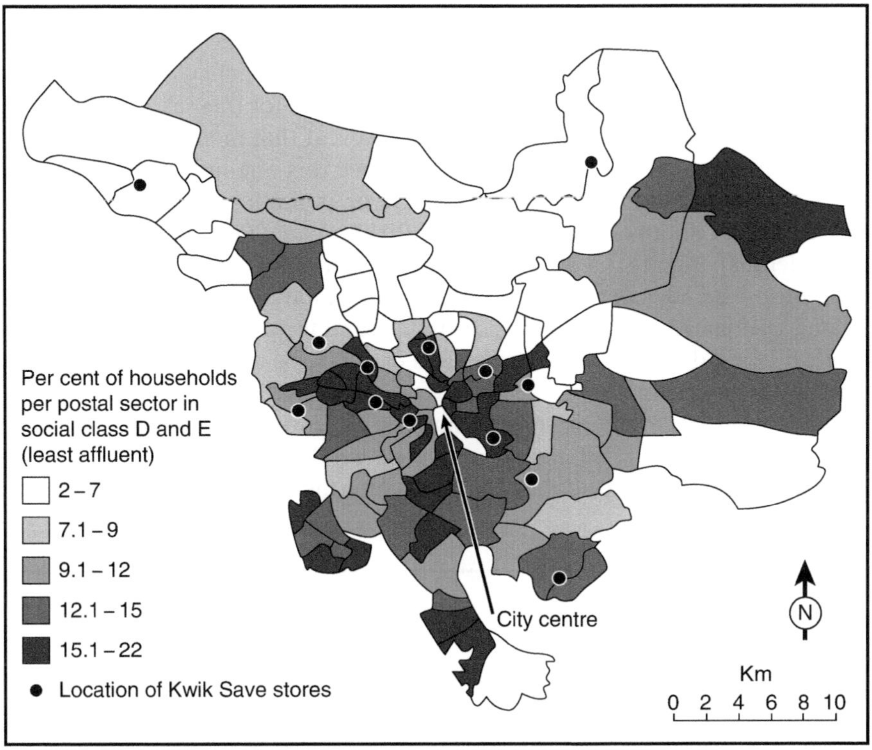

Figure 3.2 Mapping existing discount stores against low affluence.
Source: Adapted from Benoit and Clarke (1997)

by technology which allows three-dimensional mapping and animation. Hernandez (2007) provides a good review of these developments which he terms geovisualisation. The use of 3D mapping for example allows retailers to plot not only store locations across a city but also to represent size or turnover by height (a facility only available in the past in specialist graphics packages). Animation can also be useful for plotting time series data. It is difficult to convey the power of animation in 2D documents such as this. However, the reader can try various websites to see animation in practice – one of the best relates to the growth of Walmart across the USA seen in http://projects.flowingdata.com/walmart/.

3.3 Buffer and overlay analysis

GIS, however, is not simply about mapping. In most standard GIS packages there are spatial analysis routines that allow data to be manipulated and analysed in new ways. The buffer procedure is standard to most GIS packages.

Normally this involves the user drawing a distance or travel time boundary around a retail store (10 miles or 10-minute drive-time for example: again also see Section 3.4). For retailers the most common use of the buffer procedure is to demarcate the catchment area of existing or potential new stores. The first step might be to estimate the typical catchment area for their existing stores in the chain. For example, this analysis might reveal that most customers come from within a one-mile catchment area. Hence this one-mile radius can now be used in determining the size and nature of potential catchment areas for new stores (in an analogue fashion: see Birkin *et al.* 2002 and Chapter 5 for more details). Figure 3.3 shows three buffers drawn around a potential new store in south Leeds in the UK (in this case the analysis is made more sophisticated by demarcating a primary, secondary and tertiary buffer). The population or demand resident in these buffers can now be estimated using overlay techniques (see below).

Note, however, that drive-time buffers are more commonly used than straight-line buffers. The area demarcated by the buffer in Figure 3.3 (let us take the large tertiary buffer for illustration) is made up of several postal

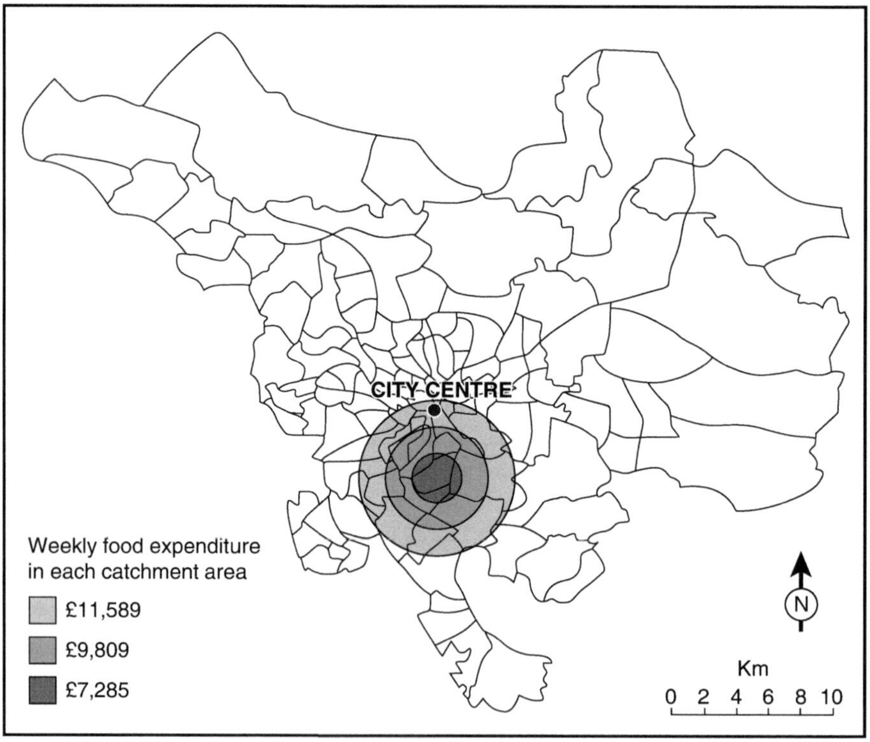

Figure 3.3 Drawing buffers around a potential new store in south Leeds (£000s).
Source: Adapted from Benoit and Clarke (1997)

sectors (the polygons shown on the map). So a key question is how can the GIS be used now to estimate the demand which lies within this buffer? Postal sectors that fall entirely within the buffer are easy to deal with – we can simply take the total population in each zone and sum to give a starting figure. However, towards the edge of the buffer the postal sectors are only partly included within it. Figure 3.4 illustrates this spatially. At its edge the buffer cuts into many postal sectors and hence the task is to estimate the population that might fall into these divided or split zones. The problem of estimating demand within these split zones is called interpolation. Figure 3.4 helps to explain and understand interpolation better.

In Figure 3.4 the buffer around a new store (called the study area buffer here) cuts an example census zone at its eastern end. The census zone contains 10,000 people. The study area buffer includes approximately 10 per cent of the area of the census zone and hence the standard interpolation procedure would suggest we need to add 1,000 people into the study area buffer population estimation. The problem with this procedure is that the population may not be evenly distributed across the census tract thus making the estimation of the population contained with the buffer problematic. If, for example, the population is actually clustered in the western area of the zone, the population living within the area cut by the study area buffer might be very small or even zero. Alternatively, if the population is concentrated in the eastern area of the census zone then perhaps most of the 10,000 people should realistically be allocated to the study area buffer. The most advanced GIS packages offer more sophisticated interpolation techniques but these are seldom seen in retail applications (see Oliver and Webster 1990; Flowerdew *et al.* 1991; Li and Revesz 2004 for more examples of alternative techniques for interpolation).

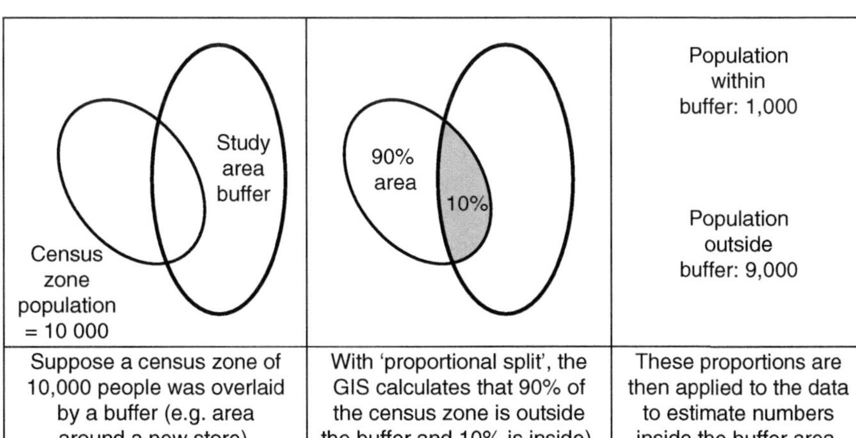

Figure 3.4 Interpolation procedure within GIS.
Source: Authors

The overlay procedure is useful in other ways. Following Benoit and Clarke (1997), we can use GIS overlay procedures to search for 'ideal zones'. Their example related to the search for a new site in Leeds, UK, for a discount retailer (building on Figure 3.2). GIS could be used as a type of sieve, systematically removing layers of information not relevant. Thus, to find a new site for a discounter we could remove all postal sectors with a high social class background (i.e. high incomes), all young persons (supposing this discounter is targeting more elderly persons) all postal sectors containing car owners and all postal sectors containing an existing (competitor) discount retailer. By overlaying these characteristics one at a time, and eliminating zones that have these features, we are left with zones that only contain older, low-income residents with no cars and no discount retailers within their locality. A population size threshold could also be added. Figure 3.5 shows the outcome of this type of sieve analysis in Leeds where a number of 'ideal zones' emerge, of which LS10 3 (to the south) seems to be the most promising. The end product shows the location of large numbers of elderly and low-income households with no cars and little or no competitor stores nearby.

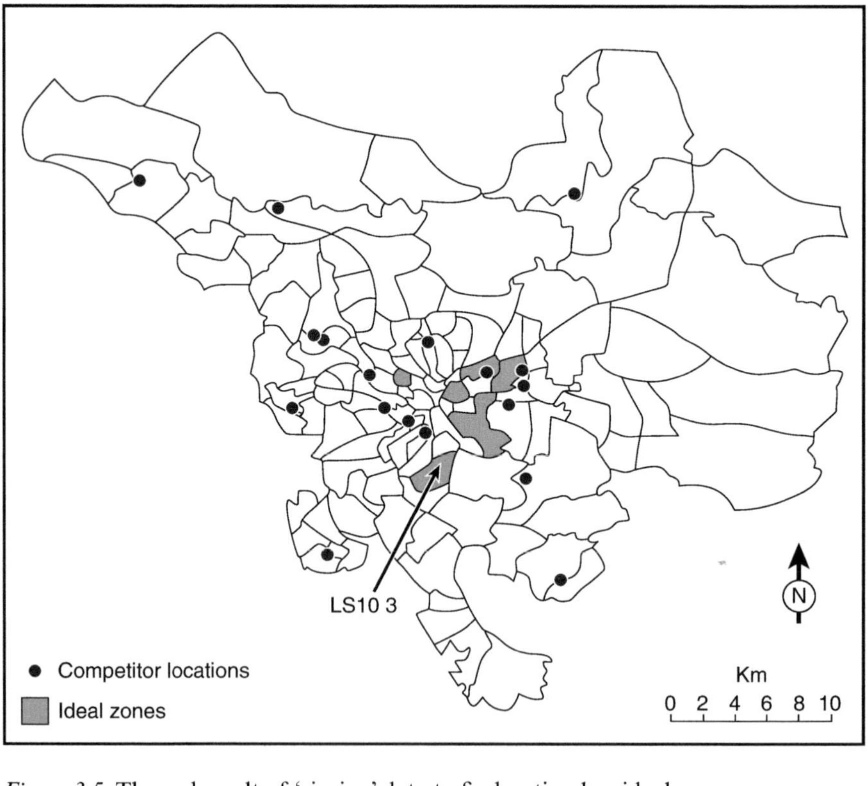

Figure 3.5 The end result of 'sieving' data to find optimal or ideal zones.
Source: Adapted from Benoit and Clarke (1997)

The so-called checklist approach utilises the overlay procedure in GIS to look at a combination of factors that might be deemed important in site location. The size or importance of a centre can be measured relatively easily by standard floor space statistics. Such aggregate statistics can be broken down by type of retailer present in order to gauge the 'quality' of that centre. In effect, the procedure is concerned with compiling as much information as possible concerning the centre and its neighbourhood. Consideration of the neighbourhood of a store would include basic population statistics drawn around such centres. These population counts could then be broken down by age, sex or social class. Thus a key question might be how many 45–60 year old persons live within a five-minute drive-time from a major shopping centre? GIS buffer and overlay procedures are again ideal techniques for this task. Comparisons of different sites would then allow the retailer to rank the possible alternatives.

Another way of overlaying or combining data is to turn the values for each variable into a standard indicator score based on a consistent scoring system. The following example is taken from the US company 'Spatial Insights' and its (undated) study of potential new locations for additional pawnbrokers or pawn shops in the city of Houston in the USA. Pawn shops are retail outlets where typically low-income residents can trade goods for cash and possibly buy them back later when resources are more plentiful (usually after pay day!). Figure 3.6 plots the current distribution of pawn shops in Houston along with a supply/demand index which, in turn, is portrayed as a 'hot spot' style indicator of demand.

This index is formed by combining or overlaying five key census variables. The first is total population. The distribution of population across Houston is divided into five equal categories. An individual census tract scores 1 point for a low population and 5 points if it falls within the top band. The second variable is household income. This time low=5 points, high=1 point, as low-income areas are more likely to provide customers for pawnbroker services. The third variable is the number of rented households (a proxy again for low social class: again low=1 point, high=5). The fourth variable is household size: large families again tend to come from poorer backgrounds in US cities. So, again low=1 point, high=5 points. Finally, population density is also measured. A high population density is also associated with lower income groups, so low=1 point, high=5 points. Thus by overlaying the scores on all five variables a very low-income area can be identified by a very high score (maximum 25). This combined score is factored again to a mark between 0 and 10. These scores are plotted on Figure 3.6 as hot spot areas and the retailers can quickly evaluate the current location of pawn shops against these high indicator scores (giving a type of index of potential). (See also the discussion of scorecards in Chapter 8.)

So, the buffer and overlay procedure seems to have a number of potentially useful applications in retail analysis. However, there are problems when this technique is used to try to estimate potential (new) store revenues. This is a

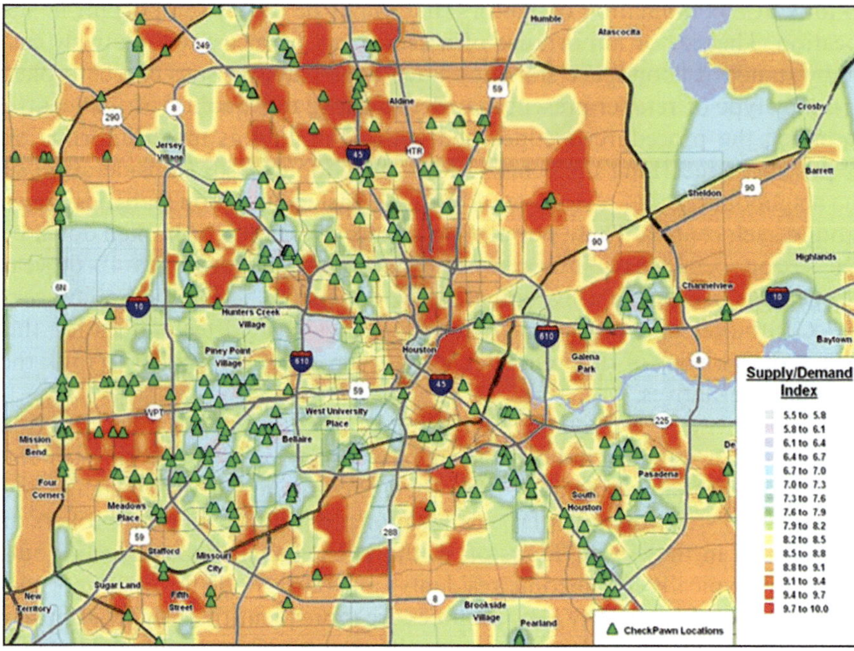

Figure 3.6 Deriving hot spots of demand for potential new pawnbrokers in Houston, USA.

Source: Spatial Insights (undated)

vital component of the store location planning exercise – yes, we can look for ideal zones or areas of high potential, but how much revenue will a new store actually attract? Let us introduce an example relating to a retailer in the USA. So far in the methodology we have shown how such retailers estimate a buffer size which can then be drawn onto a potential new site location. Figure 3.7 illustrates an example for our retailer in a typical grid-plan US city where the buffer size has been estimated at one mile. The key question is how much of the demand within the buffer, which can be estimated by overlay and interpolation, will actually end up at the new store? If there are no competitors within the one-mile boundary, the new store might be able to capture much of the demand which presently has to leave the catchment area. However, it is rare that one-mile buffers in large city regions contain no existing competitors (for most retail goods and services). According to a number of studies (i.e. Beaumont 1991) the most likely allocation procedure in more saturated markets is the fair share method. So, if three existing competitors lie within the buffer (making four in total with our retailer) then each will get approximately 25 per cent of the business. This can be factored by brand or size so that the split is not quite so even. However, it remains a fairly crude solution

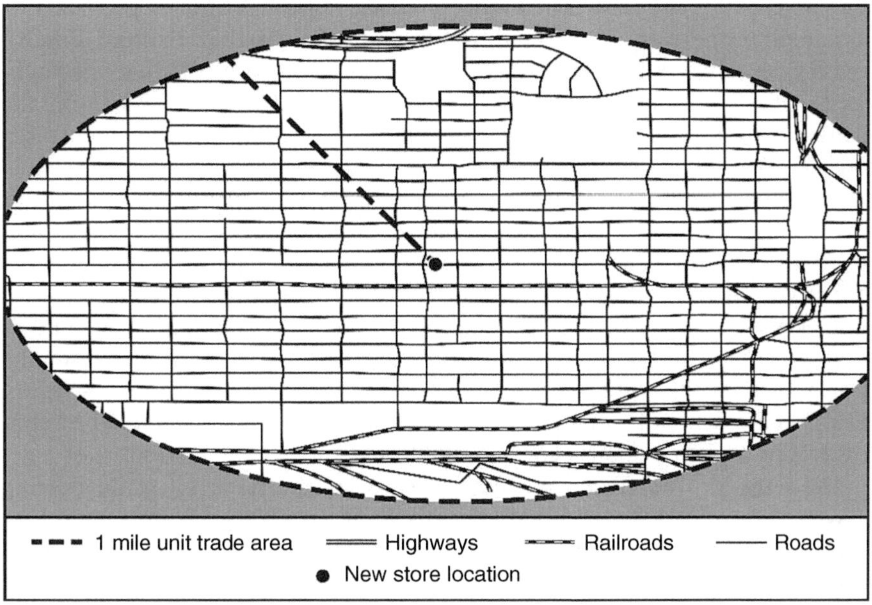

Figure 3.7 A 1-mile buffer demarcated for a new store in a typical US city.
Source: Authors

methodology. One alternative is to assume the consumer will travel to the nearest store within the catchment area (*dominant store analysis*: see Ireland 1994). However, that too is unlikely to happen in reality.

To compound the problems it is possible to think of other flaws in this technique for estimating individual store revenues. The first is the rather arbitrary nature of the one-mile buffer. Although this might work generally across the store chain, trade may be more likely to be skewed in certain directions because of the lack of competitors outside the buffer zone (irrespective of using travel time rather than straight-line distance). For example, if there were no competitors for two miles to the east of the new store location, in reality the store may get considerable trade from households outside the buffer boundary to the east (buffer inflow). Similarly, if there is a large competitor outlet just to the north of the buffer zone, perhaps the new store will get much less trade from the northern parts of the buffer as trade crosses the artificial boundary (buffer outflow). A second problem relates to the fact that there is no account taken of distance decay within the buffer. For example, the new store is actually far more likely to receive trade from nearby streets (which may or may not be very populous) and little from streets near the edge of the buffer. Hence the assumption of fair share allocation from within the whole buffer is also likely to be problematic. This problem can be reduced by identifying a primary, secondary and tertiary catchment area within the one-mile buffer (as shown in Figure 3.2) but there will still be problems of assigning proportions of demand from each of these to the new store. Finally, in very densely populated areas with lots of competitors, the individual catchment areas will become so blurred as to make any attempts of allocating the demand appropriately almost impossible (the 'overlapping catchment area problem').

A potential solution to this problem is to try to identify a unique buffer for every store. This will mean some having very small buffer sizes and others very large (depending on the amount of competition nearby). An effective way to do this is to use Thiessen polygons, also known as Voronoi polygons. Thiessen polygons can be constructed in a GIS so that each polygon contains exactly one of the stores such that any location within the polygon is closer to that store than to a store within any other polygon. Using Thiessen polygons as trade areas, analysts can evaluate the retail outlets in terms of socio-economic and/or demographic attributes of each outlet's trade area (Pearson 2007). However, the issue of trade flowing across Thiessen polygon boundaries is huge and again doubts might be given for using this technique to accurately forecast retail sales. Figure 3.8 shows the plotting of Thiessen polygons for two rival grocery stores in a region of the USA.

So, for the reasons discussed above, it is important to consider alternative methods for estimating store revenues. This will be undertaken in Chapter 5.

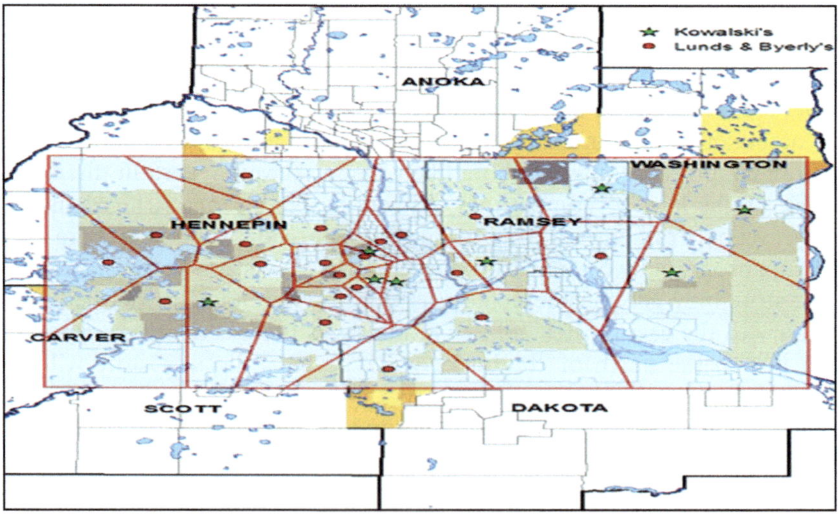

Figure 3.8 Use of Thiessen polygons for trade area demarcation.
Source: Pearson (2007)

3.4 GIS for network analysis

A more comprehensive analysis of the road network, in relation to local population distributions and the supply of shops, allows retailers to explore issues related to accessibility and provision as well as to estimate more accurately the time taken to travel (either by consumers to shops or by retailers to stores from distribution depots). Most GIS packages, for example, allow the user to explore a variety of functions for spatial analysis relating to road networks. Shortest path routines are commonplace and can be made more realistic by adding one way systems, turns, nodes and obstructions. In particular, many researchers have used such networks to identify the quickest routes from origin I to destination J in order to estimate accessibility to either single stores or entire store networks.

Networks in GIS can also be used to find the nearest store for consumers in any given region. It is thus possible to add new facilities and re-allocate demand on a nearest store basis or plan an entire network on the basis of minimising the distance individual consumers would need to travel. To undertake this exercise, many sophisticated GIS packages offer *location-allocation models*. Location-allocation models have a rich history of applications in human geography and planning (see Church and Revelle 1974; Church and Sorensen 1996; Murray 2010; Tomintz *et al.* 2015 for broader introductions and illustrations of applications). The broad aim of these models is to find the optimal location of {n} facilities given the unequal spread of demand across a city or region. The models first locate the stores optimally in relation to that demand and then they allocate the demand to each facility based on shortest distance travelled. The most common methodology has been the 'p-median model'. To solve the p-median problem the following data is required:

- Number of demand sites (i.e. census tracts).
- Number of required supply sites (i.e. potential retail locations).
- Distance, time or cost of travel from each demand site to each potential supply site.
- Number of facilities to open (i.e. number of stores required).

The p-median problem can be written as follows:

Objective function:

$$\text{Minimise } Z = \Sigma_{i \in I} \Sigma_{j \in J} a_i d_{ij} x_{ij} \qquad (3.1)$$

Subject to the constraints:

An individual demand site must be assigned to a facility $x_{ij} \leq x_{jj}$ for all (i, j)
Demand must be assigned to an open facility $\Sigma_{j \in J} x_{ij} = 1$ for all i
Exactly *p* facilities must be located $\Sigma_{j \in J} x_{jj} = p$ for all j

All demand from an individual demand site is assigned to only one facility $x_{ij} = (0, 1)$ for all (i, j)

Where:

Z is the objective function
I is the set of demand areas and the subscript i is and index denoting a particular demand area
J is the set of candidate facility sites and the subscript j is an index denoting a particular facility site
a_i is the number of people at demand site i
d_{ij} is the distance or time (travel cost) separating place i from candidate facility site j
x_{ij} is 1 if demand at place i is assigned to a facility opened at site j or 0 if demand at place i is not assigned to that site; p is the number of facilities to be located

These models have been most successfully applied to public sector location planning, for example in the location of fire stations, schools, hospitals etc. These types of services are generally less likely to be competitive and therefore can be centrally planned by a variety of planning agencies and service providers to minimise total distance travelled and hence to maximise accessibility (or to cover the most people in a predetermined distance from each service location). In the private sector the use of such models is more problematic since retailers are interested in maximising revenue not accessibility to consumers. Thus retailers starting in optimal locations for the consumer may end up clustered together as they try to steal more territory from their neighbours (this process was depicted in Hotelling's classic model of retail competition in 1929, often illustrated by other researchers through the example of ice cream men on a linear beach, although Hotelling himself never used this example: see Figure 3.9).

In the beach scenario two ice cream sales persons start selling at locations x and y (in Figure 3.9), which are the optimal locations for the consumers. However, the seller at y realises that he/she can steal more sales by moving closer to x and having a larger share of the beach. The vendor at x realises what is going on and moves closer to y. The equilibrium point is then $\frac{1}{2}(x+y)$ at the centre of the beach. Location-allocation models would wish to locate sellers at x and y not $\frac{1}{2}(x+y)$. This theory has been used as one explanation (in part) for the development of shopping centres or clusters of comparison-style retailers.

That said, some types of retailing may be more suited to the use of location-allocation models. For example, chemists or pharmacies, post offices and lottery agencies may well be more suited to the principle of maximising consumer accessibility, and indeed, are often the least clustered of all retail

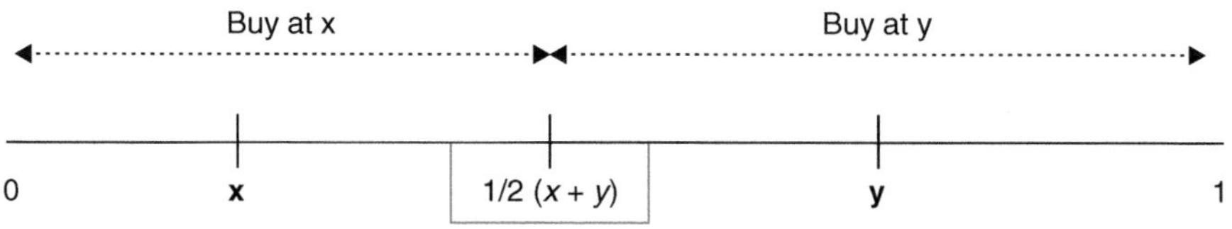

Figure 3.9 Illustration of Hotelling's theorem (1929).
Source: Redrawn by authors

Table 3.2 Average distance from major demand points in Hong Kong.

Potential location	*Total distance (km)*	*Average distance (km)*
Aberdeen	258	11.73
Central	210	9.54
Causeway Bay	219	9.96
Kennedy Town	232	10.54
Tsimshatsui	192	8.72
Tseung Kwan O	256	11.64
Tai Po	340	15.47
Tuen Mun	455	20.70

Source: Yu *et al.* (2007)

activities. A number of researchers have tried to make the location-allocation model more applicable in a competitive retail environment (Ghosh and Craig 1984; Goodchild 1984; Ghosh and Harche 1993; Farhan and Murray 2006). Yu *et al.* (2007), for example, provide an example of locating shopping malls via location-allocation models within a GIS framework. They explore four different kinds of location-allocation models to find the optimal locations for shopping malls in Hong Kong. Table 3.2 shows the types of results possible. Of the eight possible locations tried in Hong Kong, Tsimshatsui is the optimal location in terms of minimising the distance travelled for residents across the whole of Hong Kong.

Finally, network analysis could be used to evaluate alternative retail distribution systems (i.e. the locations of warehouses or depots to serve existing or new store networks). This is important for two reasons. First, retailers wish to minimise transport costs whenever possible. Second, there is increasing concern over food miles, the distance food travels from source to shop or home. Greater food miles normally equates to greater carbon emissions. In addition, shorter food miles imply more local food sources are being used. This can appeal to consumers keen to support local farms and food manufacturers. Hughes *et al.* (2017) show how it is possible to use GIS to estimate the impacts of alternative distribution systems. These studies

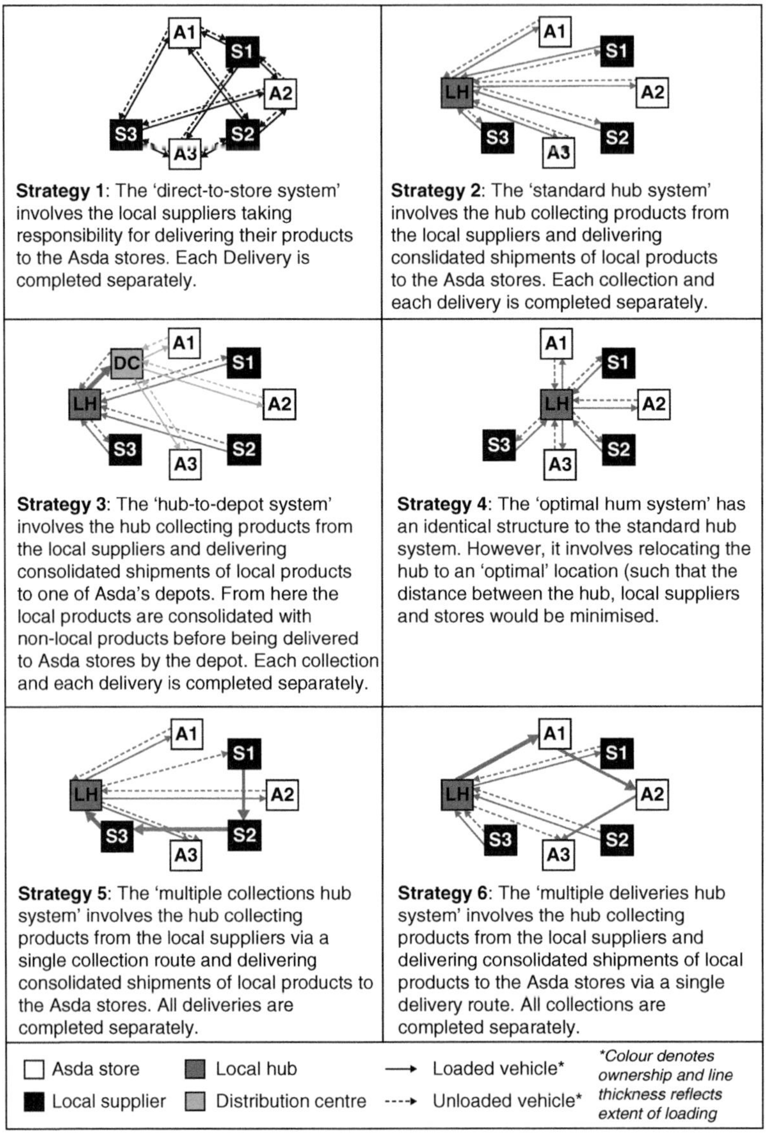

Figure 3.10 Six alternative distribution strategies for local supplies to Asda stores. Source: Hughes *et al.* (2017)

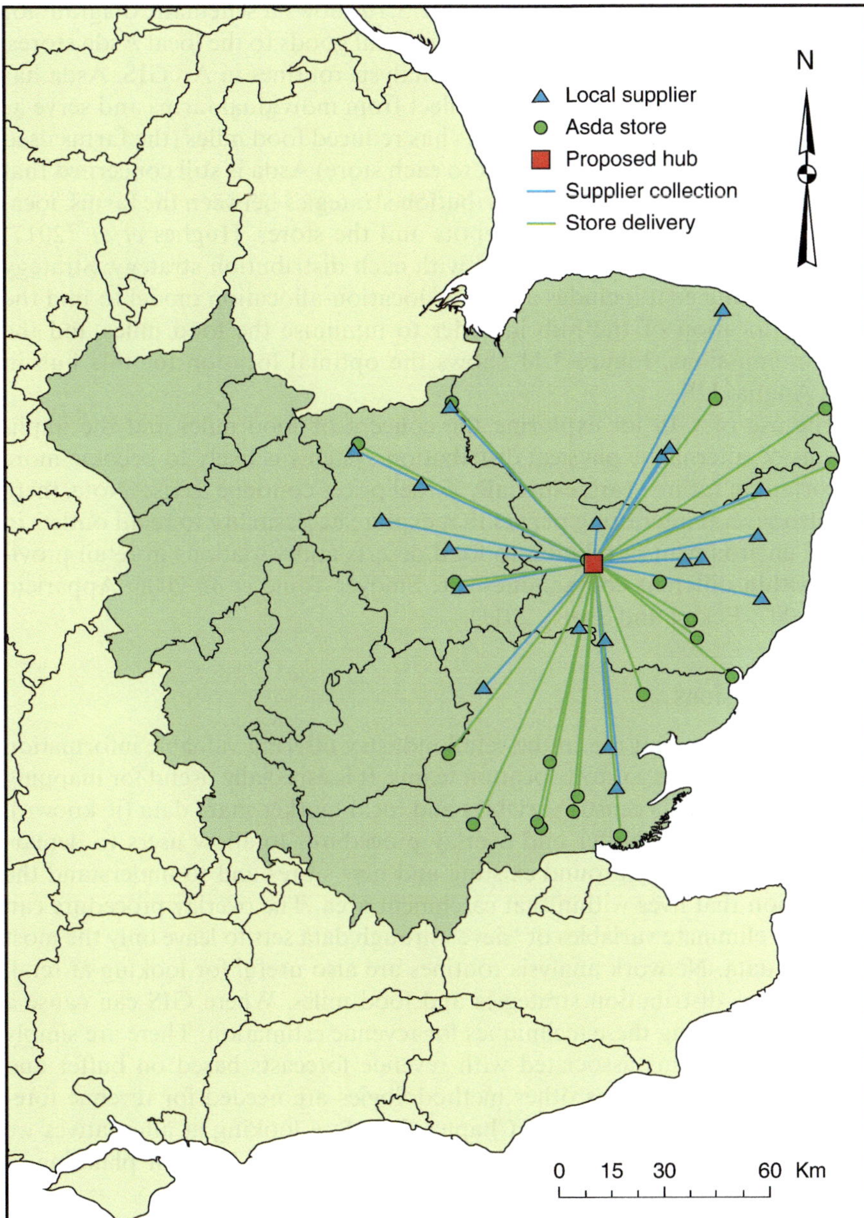

Figure 3.11 Optimising the location of the local hub to minimise the food miles associated with the distribution of local foods, East Anglia, UK.

Source: Hughes *et al.* (2017)

explore a number of possibilities for distributing locally sourced food to Asda stores in eastern England. Figure 3.10 shows a schematic diagram for six alternative scenarios for distributing local goods to the local Asda stores, each undertaken using GIS network analysis routines in ArcGIS. Asda has a unique set of local hubs, which collect from individual farms and serve as wholesalers for Asda. Although this has reduced food miles (the farms used to make individual journeys direct to each store) Asda is still concerned that there might be more optimal distribution strategies between the farms, local hubs, other Asda distribution depots and the stores. Hughes *et al.* (2017) examine the food miles associated with each distribution strategy. Strategy 4 is interesting as it includes a type of location-allocation model to find the optimal location of the hub in order to minimise the food miles and the carbon emissions. Figure 3.11 shows the optimal location for this hub in East Anglia, UK.

The use of GIS for exploring the concept of food miles and the implications of alternative physical distribution systems is likely to become more important in future years, especially if fuel prices continue to rise. Note, there are also several interesting papers that explore accessibility to retail outlets in a GIS environment in relation to food deserts and variations in retail provision within different communities (i.e. Smoyer-Tonic *et al.* 2006; Apparicio *et al.* 2007; Eckert and Shetty 2011).

3.5 Conclusions

GIS is in widespread use in the retail industry offering valuable information and spatial analysis to store location teams. It is especially useful for mapping spatial data such as census variables and local market share data (if known). It can also provide buffer and overlay procedures to allow users to demarcate catchment areas around existing and new stores and to understand the population that lives within that catchment area. The overlay procedure can also help eliminate variables or 'sieve' through data sets to leave only the most relevant data. Network analysis routines are also useful for looking at retail accessibility, distribution strategies and food miles. Where GIS can cause a problem is in using these techniques for revenue estimation. There are simply too many problems associated with revenue forecasts based on buffer and overlay procedures. Thus other methodologies are needed for revenue forecasting. These are explored in Chapter 5. Before looking at alternatives we explore the importance of geodemographics in retail location planning, a technique highly synonymous with the use of GIS.

4 Geodemographics and its role in retail marketing and location planning

4.1 Introduction

The expression 'geodemographics' has a literal interpretation as spatial or geographical variation in the characteristics of the population (i.e. demographics). It also has a slightly more focused interpretation as the presentation of single indicators of neighbourhood-level variations, and the associated classification systems, with names like Mosaic and Acorn, to which they give rise. It is the latter definition which is explored in this chapter. An understanding of geodemographic systems and their application within retail planning is important in view of the global reach of the technique and associated products (see Section 4.5). Although it is not straightforward, or necessarily helpful, to try to quantify the market for geodemographics, it is clear that most of the world's major retailers are using one of the products discussed in this chapter as an integral part of their business planning processes. The approach is also widely used among local authorities and public sector service providers (Longley 2005).

This chapter begins with a short overview of the history of geodemographics in Section 4.2. From the early beginnings of Charles Booth's London poverty maps from the nineteenth century, the development of factorial ecology techniques will be discussed as a means for the description and classification of neighbourhood geographies, first in US cities and later in the UK. Retail markets are strongly influenced by geodemographics. Coming up to the present day, we also discuss the ubiquity of geodemographics as an analytic device.

In Section 4.3 we introduce the 'Output Area Classification' (OAC) derived from the UK census, a system built by academics rather than commercial organisations and thus more open to scrutiny. A brief overview of the process of data selection, the classification methods employed and the practical application of the OAC will be described in order to give the reader a good understanding of the fundamentals of geodemographic analysis. Some of the strengths and weaknesses of OAC (*vis-à-vis* commercial classification systems) will also be investigated. Section 4.4 will be used for a discussion of various applications of geodemographics to problems of business and service planning. We will argue that it is important to begin here with advertising and direct marketing given the historical significance of work in this area. From there, the discussion progresses to geodemographics and location planning, which are closest to the core interests of this book. The importance of geodemographics for public service planning, in particular regarding questions of resource allocation, is also briefly discussed. In Section 4.5 we briefly review international examples of geodemographics. In Section 4.6 we address the question of robustness. The enduring attraction of the classifications will be considered in light of their intuitive appeal, their sound foundation in spatial processes, and the reliability in which the representation of real world patterns is conveyed. We will also assess the weaknesses of geodemographics. These stretch from questions of methodology and technique, across to ethical dilemmas, and into more practical issues regarding data and effectiveness. In the concluding part of the chapter, we will provide some reflections and a view of likely future trends and directions. A defence of geodemographics will be mounted against the invasion of 'micro' classifications founded on increasingly widespread individual-level data.

4.2 The history of geodemographics

The work of Charles Booth on the distribution of poverty in London has become quite well known. Booth used the results of surveys conducted in the central area of London at the end of the nineteenth century to characterise the affluence of neighbourhoods through a scale of seven groups ranging from 'wealthy' to 'vicious, semi-criminal'. A fuller description and online versions of the maps can be found at the Charles Booth online archive (Booth, undated). Although the methodology and data which underpin Booth's maps have relatively little to do with contemporary geodemographics, this work is notable as the first attempt to describe and classify the spatial structure of a major urban area. In passing, it is also worth noting many similarities in the patterns of deprivation and relative advantage between the 1890s and the present day (Orford *et al.* 2002). The robustness and relative stability of neighbourhood classifications over time is an enduring part of their appeal on which we will comment further below. The ideological content to Booth's description of the groups has also attracted some comment in the literature – another of the groups is labelled as 'chronic want'! Again the echoes of the debate about the prejudicial nature of group names and descriptions still resonate in the present day, as we shall also discuss later.

More sustained efforts to theorise and then analyse the spatial structure of urban areas were undertaken in the USA in the first part of the twentieth century. The work of the social ecologists in Chicago in the 1920s (i.e. Park and Burgess 1925) is particularly well known. This research has been widely discussed as the basis for the full range of contemporary models of urban form (see for example Pacione 2005 although similar discussions can be found in many textbooks of urban geography). What has attracted slightly less comment is that these attempts to model structure inevitably led to attempts to measure or classify neighbourhood composition in the city. The techniques to achieve this involve the use of small-area census data to try to distinguish the major components of spatial variation using the statistical technique of factor analysis – hence the approach became known as factorial ecology (Shevky and Bell 1955), with early applications to Los Angeles (Shevky and Williams 1949) and San Francisco (Bell 1953). By emphasising the importance of factors such as life stage, affluence and ethnicity, factorial ecology was able to demonstrate a basis in evidence for the social ecology of cities.

Attempts to transfer the methods and philosophy of factorial ecology from the USA to the UK led directly to the formation of the first geodemographic classifications. This work was stimulated by the availability of appropriate data following the 1951 Census of Population, and by a growing awareness and enthusiasm for analytical techniques during the 'quantitative revolution' in geography in the late 1950s and 1960s (Johnston and Sidaway 2004). The work of Moser and Scott (1961) to classify 157 towns in England and Wales using 60 census variables is a notable and representative example. Related work at this time, in cities such as Liverpool, Manchester and London, begins to focus on the classification of small areas at the intra-urban scale and the use of such classifications as a means for assessing deprivation and providing a basis for the effective distribution of social service resources (e.g. Edwards 1975). While such studies were typically undertaken at the scale of an individual city – partly because of the complexity of the data analysis tasks at that time – the pioneering work of Richard Webber led soon enough to the creation of a first national classification to allow the harmonious comparison of small areas across all of Britain's towns and cities. From this work, Acorn – A Classification of Residential Neighbourhoods – was born.

Thus we can see that the factor analysis methodology – including the closely related techniques of principal components analysis and cluster analysis – provided a basis for widespread studies in both the UK and USA targeted towards an academic and intellectual understanding of urban form, closely linked to the study of deprivation with associated social and policy goals. Soon after the creation of the first national system, however, researchers quickly appreciated the value of small-area classifications in discriminating not just levels of deprivation between neighbourhoods, but in assessing the behaviour and purchasing characteristics of residents in those areas. In classic work of the late 1970s, Bermingham *et al.* (1979) combined spatial data from the Acorn classification with survey data from the British Market Research Bureau and found that the level of discrimination provided by the neighbourhood classification was actually more effective than individual-level discriminators such as age, education and social class.

A good example of a geodemographic package in wide use across the UK is CACI's Acorn system. In the prosperous groups (A–C) propensities are high for expensive holidays, exclusive leisure pursuits and purchasing the quality newspapers; in contrast to the somewhat more prosaic habits and preferences of the less advantaged. With the appearance of Acorn, it is fair to say that the discipline of geodemographics – a universally applicable system for the classification of neighbourhoods in towns, cities and rural areas – was truly born. The commercial significance of the technique gave rise to many applications in target marketing and retail planning, as we will discuss further later on, and this allowed investment in the progressive updating and refinement of the systems. In the early 1980s, Mosaic quickly appeared in the UK as a competitor to Acorn, and despite the appearance of many later alternatives these remain the dominant products in the UK marketplace today. In the USA, a similar position in the marketplace is occupied by companies such as Claritas, Polk, Experian and Compusearch.

4.3 The output area classification

As we observed earlier, many proprietary geodemographic classifications, such as Acorn, Mosaic, Prizm, Cameo and others are widely used by retail and service organisations. Some insights into the creation and deployment of such classifications are provided in the literature (Harris *et al.* 2005 is probably the best available); nevertheless their commercial value means that the major detail concerning the construction and deployment of these classifications remains a carefully guarded secret. Indeed, although it is commonly understood (and of central importance) that geodemographic classifications provide a degree of explanatory and predictive value in the assessment of distributions of consumption, expenditure and product ownership, examples are poorly documented to the extent that the value added is often difficult to quantify. Although many systems are now freely available for academic use (e.g. academic licences for the use of Mosaic have been available to the academic community in the UK at least since the 2001 Census) the data can usually only be obtained at a level of aggregation that precludes investigation of the most interesting questions (e.g. in the above example, Mosaic profiles are available for rather large postal sectors, rather than for more useful output areas or individual postcodes). It is hard to escape the conclusion that detailed and critical intellectual examination of alternative systems is simply not in the business interest of these data providers.

The Output Area Classification (OAC) for the UK has been developed as a joint doctoral research project between the UK Office for National Statistics and the University of Leeds following publication of data from the 2001 Census (a 2011 version is also now available). OAC aims to provide a classification system that is freely available for use in commercial or public policy applications, and even more importantly provides an 'open' view of the data and methods used in its creation (Vickers 2006; Vickers and Rees 2007; see also the updated version for the 2011 Census produced by the Office of National Statistics (ONS) and a team at the University College London). For any geodemographic system, a necessary starting point is to identify the sources of data to be used in the construction process. The creation of the OAC begins with an assessment of the composition of established products, as shown in Figures 4.1 and 4.2.

First, in Figure 4.1, we simply consider the number of variables used in different classifications identified by Vickers (2006). At that time, the commercial classification Mosaic used the largest number of variables at 137, although a secondary classification of households used less than 60. Super Profiles and GB Profiles were both developed through academic research and Super Profiles was later exploited commercially by CDMS (Brown 1988; Brown

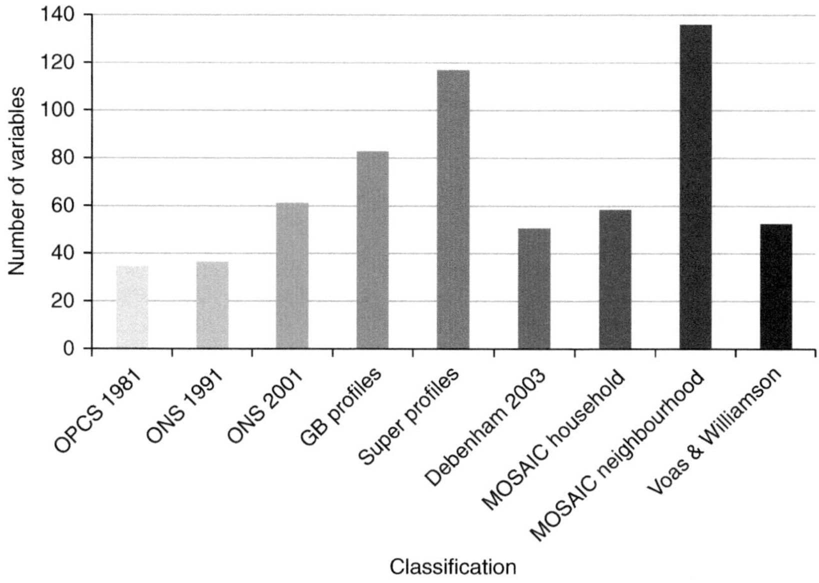

Figure 4.1 Classification detail for a range of geodemographics examples (after Vickers 2006)

and Batey 1994). The socio-economic classification of Voas and Williamson (2000) was produced for the purpose of the investigation of aggregation and ecological effects in the classification process. The work of Debenham *et al.* (2003a, 2003b) concentrates on the significance of supply-side and interaction variables alongside traditional consumption and demand-based indicators. Previous efforts by the UK Office for National Statistics are somewhat more narrowly defined with fewer than 40 variables.

The components included in the OAC are shown in Table 4.1 and comprise a relatively restricted set of 41 variables. The main reason for this is that although a wide variety of candidate variables were considered, these were systematically assessed and many rejected on the grounds of redundancy, inconsistency, unreliability or simply lack of value (Vickers 2006: Chapter 5). Ample variety is provided with the inclusion of elements representing demographics and ethnicity, housing and household composition, socio-economic status and employment. Reasonable proxies for health and accessibility are also included with journey-to-work indicators and long-term limited illness.

While the data types provided within the census are extensive, they are by no means exhaustive. It is now common practice within many commercial systems to include other indicators of the social composition of areas including wealth and expenditure patterns and consumer behaviour. Examples include shareholder registers, County Court Judgments (i.e. measures of financial default at a neighbourhood level) and a barrage of behavioural indicators from lifestyle surveys which may embrace newspaper readership, leisure

Table 4.1 The data mix for geodemographic classifications.

Demographics: Age	
v1	Age 0–4: *Percentage of resident population aged 0–4*
v2	Age 5–14: *Percentage of resident population aged 5–14*
v3	Age 25–44: *Percentage of resident population aged 25–44*
v4	Age 45–64: *Percentage of resident population aged 45–64*
v5	Age 65+: *Percentage of resident population aged 65+*
Demographics: Ethnicity	
v6	Indian, Pakistani or Bangladeshi: *Percentage of people identifying as Indian, Pakistani or Bangladeshi*
v7	Black African, Black Caribbean or Other Black: *Percentage of people identifying as Black African, Black Caribbean or Other Black*
v8	Born outside UK: *Percentage of people not born in the UK*
Demographics: Population density	
v9	Population density: *Population density (number of people per hectare)*
Household composition	
v10	Separated/divorced: *Percentage of residents 16+ who are not living in a couple and are separated/divorced*
v11	Single-person household (not pensioner): *Percentage of households with one person who is not a pensioner*
v12	Single-pensioner household: *Percentage of households which are single-pensioner households*
v13	Lone-parent household: *Percentage of households which are lone-parent households with dependent children*
v14	Two adults no children: *Percentage of households which are cohabiting or married couple households with no children*
v15	Households with non-dependent children: *Percentage of households comprising one family and no others with non-dependent children living with their parents*
Housing	
v16	Rent (public): *Percentage of households that are public sector rented accommodation*
v17	Rent (private): *Percent of households that are private/other rented accommodation*
v18	Terraced housing: *Percentage of all household spaces which are terraced*
v19	Detached housing: *Percentage of all household spaces which are detached*
v20	All flats: *Percentage of households which are flats*
v21	No central heating: *Percentage of occupied household spaces without central heating*
v22	Average house size: *Average house size (rooms per household)*
v23	People per room: *The average number of people per room*
Socio-economic	
v24	HE qualification: *Percentage of people aged 16–74 with a higher education qualification*
v25	Routine/semi-routine occupation: *Percentage of people aged 16–74 in employment working in routine or semi-routine occupations*
v26	2+ car household: *Percentage of households with two or more cars*
v27	Public transport to work: *Percentage of people aged 16–74 in employment usually travel to work by public transport*

Table 4.1 cont.

v28	Work from home: *Percentage of people aged 16–74 in employment who work mainly from home*
v29	LLTI (SIR): *Percentage of people who reported suffering from a limiting long-term illness (Standardised Illness Ratio, standardised by age)*
v30	Provide unpaid care: *Percentage of people who provide unpaid care*
Employment	
v31	Students (full-time): *Percentage of people aged 16–74 who are students*
v32	Unemployed: *Percentage of economically active people aged 16–74 who are unemployed*
v33	Working part-time: *Percentage of economically active people aged 16–74 who work part-time*
v34	Economically inactive looking after family: *Percentage of economically inactive people aged 16–74 who are looking after the home*
v35	Agriculture/fishing employment: *Percentage of all people aged 16–74 in employment working in agriculture and fishing*
v36	Mining/quarrying/construction employment: *Percentage of all people aged 16–74 in employment working in mining, quarrying and construction*
v37	Manufacturing employment: *Percentage of all people aged 16–74 in employment working in manufacturing*
v38	Hotel and catering employment: *Percentage of all people aged 16–74 in employment working in hotel and catering*
v39	Health and social work employment: *Percentage of all people aged 16–74 in employment working in health and social work*
v40	Financial intermediation employment: *Percentage of all people aged 16–74 in employment working in financial intermediation*
v41	Wholesale/retail trade employment: *Percentage of all people aged 16–74 in employment working in wholesale/retail trade*

Source: Vickers (2006)

activities, preferred retail destination, and ownership of big ticket items such as computers, mobile phones and televisions. It is significant perhaps that current views of the UK Census see at least one possible future in which a universal form is replaced by a patchwork of administrative (and perhaps commercial) data sources knitted together into a continuously updated and diverse source of social and spatial intelligence. Open source classifications of the future will need to exploit extended data sources of this type if they are to keep pace with commercial applications in which such diversity is already substantially embedded.

The next step is to think about how these systems are created. In essence, the OAC starts with a spreadsheet with 220,000 rows (representing each of the output areas in the UK) and 41 columns (one for each of the individual data items in Table 4.1). The problem is to classify or group together output areas according to shared or similar socio-demographic characteristics. While a variety of methods can be used to reach the desired outcome, the most commonly favoured approach is 'k-means', an iterative relocation algorithm in which OAs are initially grouped at random, and then progressively shuffled into more coherent groups. Vickers (2006) describes a hierarchical process in which

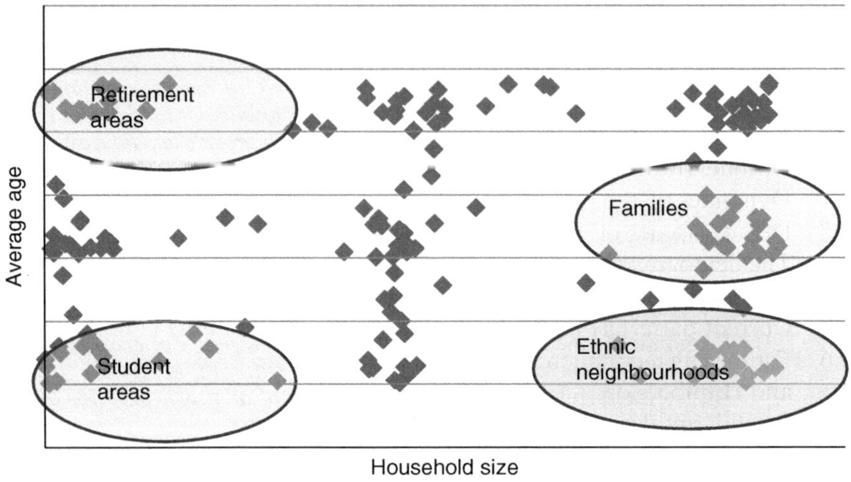

Figure 4.2 Location types in a synthetic city.
Source: Authors

the lowest level of the classification is created first, from which the centres of the clusters are then re-clustered to produce the middle level of the hierarchy. The same would be done on these to create the highest level. Similar hierarchies are generated for many geodemographic systems – for example, the version of Acorn current in October 2016 (CACI 2016) features 5 high-level 'types', 17 second-tier 'classes' and 59 low-level 'groups' (for private households).

Consider, for example, the data shown in Figure 4.2 which shows the median age and average household size for small areas in a synthetic city. Through visual inspection of this scatterplot, a series of clusters might be hypothesised – very large average household sizes could be representative of ethnically mixed areas with large numbers of extended families; other relatively young areas with reasonable household size are perhaps more likely to be family neighbourhoods in the suburbs. Smaller households might typically represent both ends of the lifecycle in demographic terms (e.g. either elderly people or young professionals living alone).

Now consider that the number and distribution of clusters is to be determined not arbitrarily but analytically for the most effective representation of inter-area differences, and that the number of indicators is increased from 2 to 41 (in accordance with Table 4.2). This is the essence of the k-means procedure – an algorithm to group areas according to common characteristics in an 'attribute space' of high dimensionality.

The final stage in the creation process is to construct profiles for each of the clusters. The key procedure here is to sum the individual cells of the data matrix by cluster. In the case of OAC, seven high-level clusters are identified and the profile of some important variables is shown in Table 4.2. If the classification process has been successful, then the profiles for each cluster will be

Table 4.2 Correlation between OAC groups and key census variables.

	Age 25–44	*Born outside UK*	*Car ownership*	*Population density*
Blue-collar communities	Average	Low	Low	High
City living	High	High	Low	High
Countryside	Low	Low	High	Very low
Prospering suburbs	Low	Low	High	Low
Constrained by circumstances	Average	Average	Low	High
Typical traits	Average	Average	Average	Average
Multicultural areas	High	Very high	Low	High

Source: Vickers (2006)

rather distinctive and may be used as a basis for naming each cluster, perhaps as a prelude to more detailed descriptions sometimes referred to as 'pen portraits' (e.g. Brown 1991). For example, in the current version of the Acorn User Guide (CACI 2016) Category 1 'Affluent Achievers' are characterised as living 'in large houses, which are usually detached' and 'a high proportion of these people are very well educated and employed in managerial and professional occupations'. These observations are clearly based on high or very high scores for variables such as degree-level education, size of house and white collar employment.

Since the clusters of the OAC, or any other geodemographic system, have been generated in order to provide maximum distinction between one another, our expectation is that similar distinctiveness will be exhibited with respect to other behavioural characteristics and consumption preferences of the population. Consider, for example, a situation in which a travel company selling holidays has a database of customers, and each of these customers is associated with an address, including a postcode. Assuming that each postcode can easily be attached to a census output area (OA), then profiles of holiday-makers can be constructed simply by attaching the relevant codes. For example, in Table 4.3, we see a selection from the customer file of the travel operator *Sunshine Holidays*. The first customer is Mr Simon Smith who lives at Number 11 West Grange, LS11 0HH. This postcode is in the output area DAFH14, and this OA has a Cameo code of 3F. The second customer is Mrs Jane Jones who lives at 14 Wesley Street, LS12 3QB. The OA is DAFH15 and the Cameo code is 3G.

The selection of records shown in Table 4.3 (in this anonymised form) represents only the first ten cases in a file of perhaps 15,000 customers. After the whole file is profiled and aggregated, the totals in Table 4.4 are generated. This becomes interesting when the customer profile is compared to the national population distributions shown in Column 4 of the same table. For example, Cameo Group 5 accounts for 23 per cent of holiday customers but only 15 per cent of the UK population. Cameo Group 8 forms 11 per cent of the UK population but only 5 per cent of the customers. One further disaggregation is shown in

Table 4.3 Variable profiles for OAC neighbourhood types.

Name	*Address*	*Postcode*	*Output area*	*Cameo*
Mr Simon Smith	11 West Grange	LS11 0HH	DAFH14	3F
Mrs Jane Jones	14 Wesley Street	LS12 3QB	DAFH15	3G
Dr Bobby Brown	25 Lakeside View	LS13 6BF	DAFJ16	4A
Ms Carole Cole	32 South Street	LS14 9KL	DAFK17	4B
Mr Timothy Taylor	45 North Road	LS15 2YR	DAFL18	4C
Mr Malcolm Marshall	2 High Street	LS16 5FV	DAFM19	5A
Mrs Katie Keegan	28 Church Road	LS17 8HS	DAFN20	5B
Rev George Green	The Vicarage	LS18 1NN	DAFP21	5C
Ms Natalie North	275 Bradford Road	LS19 4RF	DAFQ22	6A
Mr Roger Rabbitte	The Burrow	LS20 7ES	DAFR23	6B

Source: Authors

Table 4.4 Customer profiles for an imaginary data segment.

Cameo	*Customers*	*Share of customers (%)*	*Share of population (%)*	*Index*
1	23	1	4	38
2	282	18	14	134
3	358	23	20	114
4	256	17	14	118
5	351	23	15	151
6	61	4	9	45
7	104	7	8	80
8	71	5	11	41
9	31	2	4	48
Total	1,537	100	100	100

Source: Authors

Table 4.5 Cameo profiles for travel products.

Flight only		*Cruise*		*Family*	
3F	213	4G	243	5E	210
3D	207	8F	239	8C	189
2A	205	3A	231	4D	175
3E	190	4A	230	3A	172

Source: Authors

Table 4.5, in which we consider three different holiday products – family packages, flight only, and cruises. These figures are based on an average consumer profile of 100 and are based on findings with a real travel operator. Cruises are found to be most prevalent among Cameo groups 3F, 3D and 2A (typically representing retired households in wealthy areas), while for flights only the dominant groups are 4G and 8F (singles and young couples), and for family holidays then 5E and 8C are the most important (suburban family neighbourhoods).

Now that we have seen how geodemographic classifications may be generated and an example of their application, we can consider a wider range of uses and more practical examples. This follows in Section 4.4, but before moving on we will briefly consider some of the advantages of the method. First, geodemographics is intuitively plausible. The postcode for the University of Leeds shows an Acorn classification of 17 (young, educated workers, flats) while for Alwoodley Golf Club, considered to be one of the most prestigious golf clubs in Yorkshire then the code is 1 (affluent, mature professionals, large houses). All of this makes great intuitive sense. Furthermore, these findings translate into specific analytical circumstances. Thus in the previous example, it makes sense that cruises are most abundant among wealthy post-family customers, while the young and unattached are most likely to fly alone. As we saw in the earlier discussion of the first geodemographic studies of market research data (Bermingham *et al.* 1979), such profiles have consistently been found not only plausible but powerful discriminators of customer behaviour. Geodemographics are easy to understand and straightforward to apply. They are heavily reliant on census data which is high quality but detailed, cheap and accessible. It is little wonder that the technique has achieved considerable popularity.

4.4 Applications of geodemographics

If we return to our previous example, we have asserted that classification systems may be used as a basis for profiling the customer base of a retail or service organisation. This is potentially valuable in a number of regards. The core application is *response modelling*. Let us suppose that *Sunshine Holidays* wishes to promote its product through the direct marketing of brochures. A successful campaign will yield good margins, but the production and distribution of materials is expensive. Now consider the stylised case in Figure 4.3, in which the profile of customers has been ranked and cumulated. Thus 'Group 4' comprises 10 per cent of the population but yields 30 per cent of customers. Group 2 contains 20 per cent of customers among 10 per cent of the population, and so on (this can be seen by reading up the graph from the 20 per cent tick on the x-axis in Figure 4.3). In this example, therefore, 50 per cent of customers can be found in just two groups comprising 20 per cent of the overall UK population. The remaining customers are distributed among 80 per cent of the population in the other groups. If the strategy is to target customers in Groups 2 and 4 the relative chances of success are (50/20) compared to (50/80) in the remaining groups – the returns from targeting these two core groups is fully four times the expected return from the low priority groups.

A typical application of response modelling would be from a marketing trial. Before launching a full campaign, an expensive new product could be trialled in a small part of the country, or among a subset of existing customers. Based on the profile of responses to this initial exercise, the campaign is then rolled out nationally to restricted groups, perhaps based on some further model of returns and profitability.

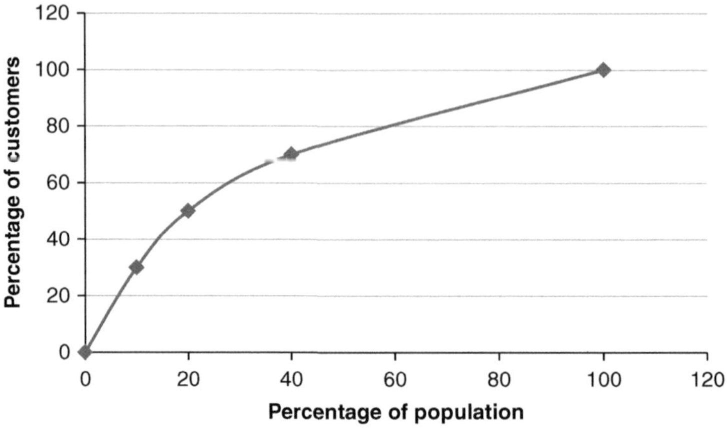

Figure 4.3 Hypothetical gains chart.
Source: Authors

The idea of response modelling has a natural extension to retail location and distribution planning. So far we have argued that the production of new brochures is expensive, and that certain population groups will respond more enthusiastically to some products than others. This being the case, then *Sunshine Holidays* will be well advised to stock up on cruise brochures in areas of 4F, 8G, 3A; while where the demographic is 3F, 3D, 4A, flights are the order of the day. If space and product are at less of a premium, it could be that it simply makes sense to display brochures or window advertising one kind more prominently than the other(s). This argument is generally applicable to retailers who will typically offer a wide variety of products through many formats. Birkin *et al.* (2003) discuss an example in which petrol retailers can provide a range of services to exploit forecourt space – for example, car wash versus quick service restaurant. It is logical to prioritise these options with respect to the profile and needs of the catchment population of different outlets.

A further extension of the same idea could suggest that geodemographic profiling is viable for a broader range of network planning problems, such as the location of new stores. Indeed there are examples of this in the literature and in planning practice, particularly in relation to the development of new shopping centres. Birkin *et al.* (1996) discuss an example in which the potential for the major new development at Meadowhall in Sheffield, UK, is assessed through a geodemographic comparison against the customer profile of an extant scheme of a comparable scale and style at the Metro Centre in Gateshead, UK. The problem for such methods is that they do little or nothing to address the problems of catchment area definition and retail competition (see Khawaldah *et al.* 2012 for more on the definition of *retail centre* catchments). Therefore a logical conclusion is that geodemographics could provide a suitable basis for demand estimation within the retail modelling process, but not as a replacement for the technique in its entirety (see more discussion on demand estimation using geodemographics in Chapter 6).

Both the network design problem and the format optimisation problem could be characterised as a question of *resource allocation*. The retail organisation has a scarce resource – investment capital for new store development or product inventory to hold at those stores. These resources must be deployed in the most appropriate way to maximise the potential return.

For other useful reviews of geodemographics used in retail analysis see Gonzales-Benito and Gonzales-Benito (2004, 2005).

4.5 International classifications

Certainly one set of products that have aided location planning teams working in international markets has been the increased availability of global geodemographic products. Experian's EuroMosaic was probably the first, followed by products made by CACI and CallCredit (its Cameo products). Figures 4.4 and 4.5 show a good example of how these can be used to compare the socio-economic environments of different global cities using the same classifications. Both maps show the distribution of the highest income group available in Cameo's International classification, and how one can plot the same variable for different cities around the world – in this case Los Angeles and Sydney respectively.

The value of these maps for an international retailer targeting high-income consumers around the world is very obvious.

4.6 The dangers of geodemographics

We start with an example of university students to expose an area of methodological concern whenever geodemographics is discussed. Suppose we have a table of university course participation with the first row showing us that in areas of 'Wealthy executives', the participation rate for students taking medicine and dentistry is unusually high with an index of 156. Similarly, for 'Affluent greys' the rate for veterinary sciences is an even more extreme 192. Can we take this to suggest that the UK's bankers and lawyers have become bored with their wealthy executive lifestyles and are now retraining as doctors? Is it the case that businessmen who have now retired in grey-haired affluence are now considering their second careers in animal medicine? This seems unlikely: a much more plausible interpretation is that the children of these households are those with a marked interest in the medical and veterinary professions. This makes the point that distinguishing between the nature of the area and the nature of the individuals in that area, is not an entirely trivial matter.

Table 4.6 makes this point in a slightly more general way. Here we are looking at the Output Area Classifications with respect to the data which actually feeds into them. Thus, for example, if we take the age profile of areas classified as 'older blue collar', actually there are more people aged under eight than there are over 75. Conversely, in areas categorised as 'younger blue collar', there are still substantial numbers of elderly. If a travel operator wanted

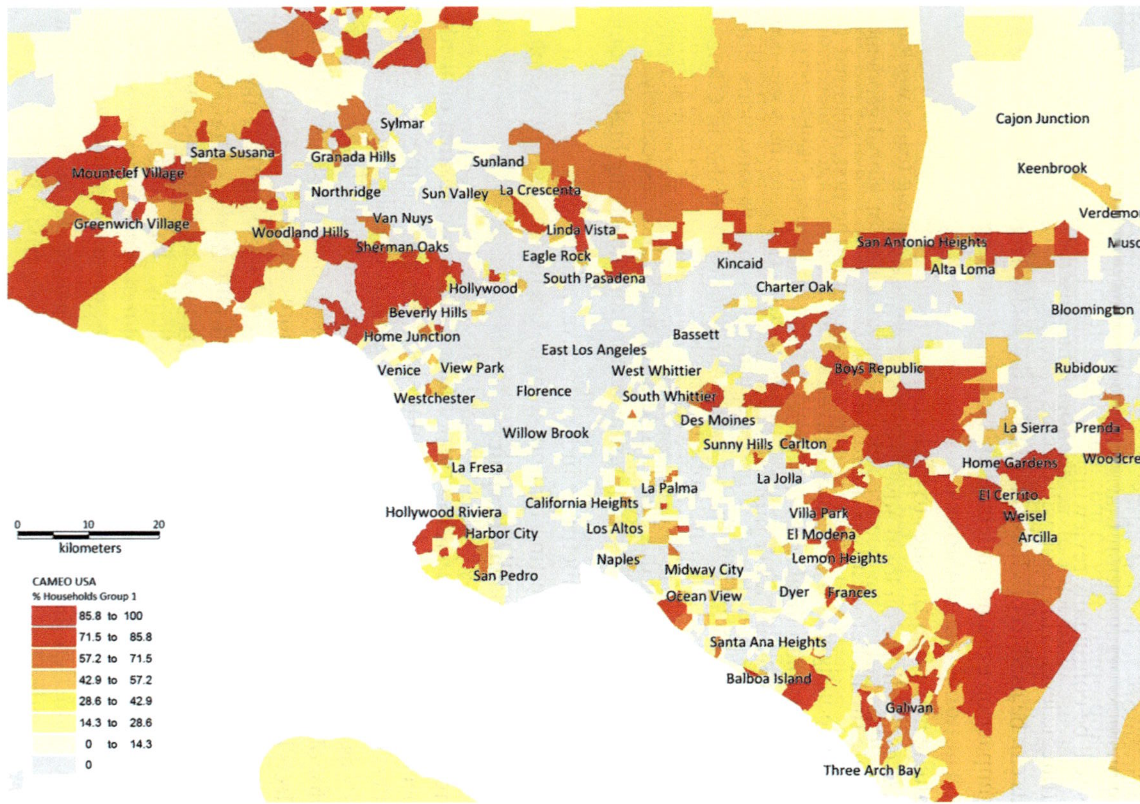

Figure 4.4 Highest income earners (Cameo group 1) in Los Angeles.

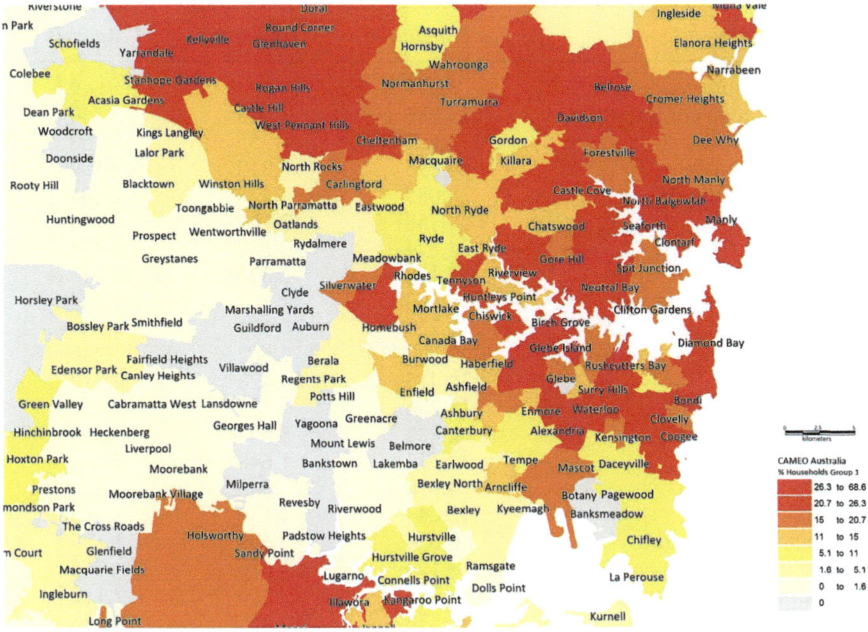

Figure 4.5 Highest income earners (Cameo group 1) in Sydney.

to target families with young children for holidays to, say, Euro Disney, a blanket mailshot to residents in younger blue collar areas will wastefully target many individuals well outside the primary customer base, while missing many others with greater potential. This problem is referred to in the literature as the ecological fallacy (e.g. Openshaw 1984) – in effect, that it is dangerous to 'average' the characteristics of any diverse group of individuals. It is manifest in many different approaches to area classification. For example, the Jarman Index (Jarman 1983) was used for many years as a basis for funding primary medical services (general practices). The index was based on a composite of census data as a proxy for deprivation (i.e. unemployment etc.). If the national average is 100, any area with a score above 130 received extra funding; any area with a lower score did not. This seems to ignore, or discriminate for, the fact that many people in deprived areas need not be considered deprived based on personal measures; while many people outside deprived areas will actually score highly on the same measures on an individual basis.

Of course, the smaller the geographical scale of analysis the more homogenous the population is likely to be and the less problematic the ecological fallacy becomes.

Finally, we should note ethical concerns over geodemographics. Some people have genuine fears that a lot, perhaps too much, is known about individuals and their lifestyles. Both Goss (1995) and Pickles (1995) were among the first to

Table 4.6 Customer penetration for selected travel products.

OAC sub-group	*Population*	*Aged 0 to 7*	*Aged 75+*
Accessible countryside	8,599	684	649
Afro-Caribbean communities	13,086	1,413	740
Agricultural	627	55	58
Asian communities	66,395	8,312	3,259
Aspiring households	28,085	2,478	1,664
Least divergent	39,748	3,169	4,264
Older blue collar	24,814	2,339	2,174
Older workers	69,158	6,325	8,218
Prospering older families	28,313	2,164	1,936
Prospering semis	72,229	6,171	4,936
Prospering younger families	20,677	2,562	330
Public housing	34,415	4,428	2,519
Senior communities	11,246	379	2,624
Settled households	62,882	5,541	3,565
Settled in the city	32,514	1,648	3,032
Terraced blue collar	4,670	505	325
Thriving suburbs	45,643	4,152	3,315
Transient communities	30,024	660	1,185
Village life	8,174	686	765
Young families in terraced homes	35,795	3,306	2,112
Younger blue collar	78,376	10,485	4,248

Source: Authors

provide major critiques of the geodemographic (and related GIS) industry on these grounds. Concerns about the increasing 'surveillance of society' and the 'commodification of people' are often raised. A full review of this debate takes us beyond the remit of this book – but those who use (and market) geodemographics should constantly think about confidentiality and individual privacy.

4.7 Discussion and future trends

Geodemographic classifications have been used over an extended period of time to provide a straightforward representation of variations in the composition of neighbourhoods in a way that is readily understood and easy to interpret. For many commercial purposes, spatial variations between consumer groups and their purchasing habits are frequently more important than individual or household variations in age, occupation and affluence. Popular systems such as Mosaic, Cameo and Acorn therefore have an important role as tools in the spatial planning armoury of major retail organisations, in which function they may have value in their own right, or as a complement to other spatial planning models such as those which we explore elsewhere in this volume (see especially Chapter 5). In the past, geodemographics have typically been developed with census data at their heart, but more recent development of the major systems has drawn on a wider corpus of measures and indicators. This trend can be expected to continue into the future, with respect to both commercial and open source geodemographic products. This direction of travel is articulated at greater length in Chapter 11 when we discuss the potential of 'big data'.

In view of the proliferation of new sources of data at ever-increasing levels of individual detail, it is necessary to consider the extent to which geodemographics may be in the process of being overtaken by individual classifications. Indeed such classifications have been available in the commercial sector for a number of years, using individual data from customer databases and various registers and directory sources. Products like Prizm Household (Claritas) and Personics (Acxiom) lead the way here. Impressive examples are documented in the literature of the way in which the UK supermarket giant Tesco has exploited data from its store loyalty card to create a classification of customers that is strongly predictive of their behaviour, expenditure patterns and future dynamics (Humby *et al.* 2008). Yet what does this tell us of those customers outside the loyalty programme, and most importantly those who are not Tesco customers at all? For marketing and promotional activity these existing customers may be of prime interest; for site planning purposes it is the unknown group who are of much greater interest.

What the glossy exterior of these classifications tends to conceal is a certain inconsistency in the underlying data. While customer databases may be rich in detail, they tend to lack reach, so that for the majority of indicators the drivers of the classification may be common denominator variables such as name and address from a registry source such as the electoral roll. However, considerable interest remains on producing more robust individual-level classifications (i.e. Burns 2014).

Geodemographic profiles are comprehensive in providing a basis for market evaluation irrespective of previous spending or loyalties, but are also much more than a convenience. As has been seen (e.g. in Section 4.3), neighbourhood typologies demonstrate a persistent ability to outperform conventional social groupings– where the individual lives is still at least as important as who they are, in demographic terms. It seems clear that geodemographic and lifestyle classifications are complementary, at the very least.

A more insidious threat to the continued good fortune of geodemographics – in the UK at least – is posed by uncertainties in the future of its underlying data sources. There seems now every chance that the next UK Census in 2021 will be the last. Nevertheless part of the logic for this possibility is the fact that many other good quality sources are now becoming widely, easily and cheaply available. If survey data could be routinely coded by geodemographic type (rather than individual address or postcode which may invoke confidentiality concerns) then they could be greatly enriched for spatial planning purposes (Birkin and G. Clarke 2012). With the right levels of imagination and ingenuity, there seems no intrinsic reason why neighbourhood classifications, perhaps with continually refreshed and diverse components, should not continue to provide robust and powerful market segmentations across local neighbourhoods into the foreseeable future and beyond.

5 Model-based methods for store network planning

5.1 Introduction

In the previous two chapters we have explored the use of GIS and geodemographics in network planning and discussed the powerful information that can be provided by both. In this chapter we recap on the main modelling techniques used by certain retailers, especially the larger corporate giants. In Section 5.2 we introduce a simple form of modelling – the analogue technique. In its basic form this compares data sets between existing and potential store catchment areas. The more advanced analogue approaches become statistical modelling techniques and involve the classic statistical methodologies of correlation and regression. We shall discuss these techniques in Section 5.3. In Section 5.4 we introduce the methodology most used and developed by ourselves over many years, often in collaboration with retail clients – gravity or spatial interaction models. Since this is the major technique explored and used as an exemplar in the rest of the book, we shall break Section 5.4 into a number of sub-themes to provide the reader with more information on how these models work.

5.2 The analogue method

Analogue techniques have always been very common procedures for site location, in the UK and USA especially. The basic approach involves attempting to forecast the potential sales of a new (or existing) store by drawing comparisons (or analogies) with other stores in the corporate chain that are alike in physical, locational and trade area circumstances. This may be done simply by examining relevant data sets or more scientifically through regression techniques (see Section 5.3). The analogue technique can operate in two ways. First, for example, imagine company X reviews the sales figures of its stores. It can rank these in many ways and already we can detect a potential problem – how do we evaluate success? We might need to think of a disaggregated form of analogue from the outset. For example, do we rank according to total sales or sales per sq ft (so that stores of different sizes can be fairly compared), or simply rank store revenue by type of store (hypermarket, convenience etc.)

or finally by type of location (large city centres, medium sized towns, edge of town green-field sites etc.).

In this case let us suppose we wish to evaluate performance across medium sized UK towns for retailer X. Suppose also that the top four sales performances for retailer X in the UK are in Harrogate, Cambridge, Lancaster and Bath. The next step might be to classify these types of centres using GIS or geodemographics. Thus it seems retailer X does best in towns classed as having 'wealthy executives' or 'inspiring singles' (perhaps a student effect in the latter) – these are two categories in CACI's Acorn geodemographic system (recall the discussion in Chapter 4). Next, the analogue technique works by finding other towns in these geodemographic categories where retailer X currently does not have a store: in this case maybe Cheltenham, Winchester (wealthy executives), Warwick and Exeter (student style towns). The priority is then to find opportunities for new store openings in these locations.

The second way the analogue technique might work is by helping to react to planning opportunities. Suppose a developer contacts retailer X and offers a site in, say, Swindon. The network planning team might then use the analogue approach to find a town similar in geodemographics to Swindon where they currently do operate a store. If that store is performing well then they are more likely to say yes to the new opportunity presented.

Does this approach work? The answer depends on whether or not it is possible to find similar sites across the country to existing locations trading well, and whether it is possible to transfer the trading characteristics across geographical locations successfully. In reality, a wide variation in performance is frequently found between outlets in a retail chain even in catchment areas with similar characteristics, perhaps because levels of competition are very different. Moreover, even if a similar geographical catchment is found to the new store, what happens if the analogous store is currently over or underperforming? For more discussion on the analogue method see Laing *et al.* (2003).

Despite these possible limitations the analogue technique can be another useful technique to help understand network performance (a short example will also be given in Chapter 9 when we consider alternatives to spatial interaction models in relation to network reorganisation).

5.3 Statistical modelling

As noted above, statistical techniques are normally associated with correlation and regression models. Regression modelling represents one of the more scientific methods of network planning and builds on the philosophy of the analogue procedure. Again, its use is popular in many retail organisations. Regression analysis works by defining a dependent variable such as store turnover and attempting to correlate this with a set of independent or explanatory variables (e.g. size of store, catchment population total, retail brand). Statistical techniques such as correlation and regression modelling are often standard features of more advanced GIS packages. They can take a number of variables stored in a GIS and test various combinations to search for good correlations between, for example, store sales data and various geodemographic data. If a good fit can be found between sales and geodemographic data for existing stores, revenues for new stores can be predicted with relative ease, at least in terms of applying the geodemographic data to new locations. Coefficients are also calculated to weight the importance of each independent variable in explaining the variation in the set of dependent variables. The model can be written as:

$$Y_i = a + b_1X_{1i} + b_2X_{2i} + b_3X_{3i} + \ldots + b_mX_{mi} \qquad (5.1)$$

where, Y_i is turnover (the dependent variable) of store i, X_{mi} are independent variables, b_m are regression coefficients estimated by calibrating against existing stores, and a is the intercept term. As there are more than one or two independent variables such a model is often called a *multiple regression model*.

If the independent variables are not known in advance then a useful procedure to adopt is *stepwise regression*. This is one of the most frequently used regression analysis methods. There is a degree of automation about this procedure. Normally, the statistical software program determines which variables among the specified set of independent variables will actually be used for the regression and in which order they will be introduced. After each step, the algorithm selects (from the remaining predictor variables) another variable to try and see if that yields a better correlation with the dependent variable. Poor performing independent variables are dropped as necessary and the model iterates until the best set of independent factors can be found. If there are many independent variables then this iterative procedure can be very powerful and ultimately time saving.

It is useful to give a few examples of the use of these models in network planning. First, Fenwick (1978) gives an example of the building of a regression model to try to correlate bank sales (for example the number of new mortgages purchased in a year) against a set of socio-economic variables. Keeping the above terminology, the best model was defined as incorporating the following variables: X_{1i} is the average age of persons in catchment area of branch i, X_{2i} is the average socio-economic status of residents in catchment area of branch i, X_{3i} is the number of years branch i has been established, while a fourth variable was the number of new houses under construction in the catchment area of branch i, and a fifth variable was the total number of banks and building societies (lenders for mortgages) in the catchment area of branch i. (note the first four are likely to positively influence branch sales while the fifth would be a negative factor – the greater the competition, the lower branch sales).

A second useful example is taken from the work of Duggal (2008). This study examined the factors which might best explain a set of revenue figures for selected McDonald's and Burger King Stores in the USA. Two regression models were built. Interestingly, ethnic population and median household income were found to be the two most important variables for both retailers. However, ethnic population and median household income explained annual sales variations for Burger King, only within buffers of one and two miles around the restaurants (see the link with GIS here to define the buffer sizes and the characteristics of the populations within those buffers). For McDonald's, ethnic population and median household income within five-mile buffers better explained the annual sales variations. This suggests that consumers seemed to be more willing to travel a longer distance to eat at McDonald's as compared to Burger King.

A third useful example comes from the work of Shields and Kures (2007). In this example the key question is whether there was a spatial pattern to store closures of Kmart in the USA following corporate financial problems. The aim of their regression model was thus to find a consistent relationship between the location of stores closed and the socio-economic or geodemographic characteristics of their catchment areas. Their model took the following form:

$$\begin{aligned} \text{open?} = \beta_0 &+ \beta_1 \text{households} + \beta_2 pct20\text{k } 50\text{k} + \beta_3 \text{wal} \\ &- \text{mart} + \beta_4 \text{target} + \beta_5 \text{other kmart} \\ &+ \beta_6 \text{transport} + \beta_7 \text{hhsize} + \beta_8 \text{poverty} + \varepsilon, \end{aligned} \quad (5.2)$$

As Shield and Kures (2007) explain, the first independent variable is the number of households while the second is income range (B used here instead of X). Variables 3,4 and 5 refer to key retailers present in the catchment area of each closed Kmart store. Variables 6,7 and 8 were again geodemographic variables relating to the size of the transport network, household size and poverty levels respectively. In addition, ε is the unobserved term normally distributed with mean=0 and variance $\sigma^2=1$ (an error term). The dependent variable takes on a value of one if the store remains opened, else it is zero. As described, the model is a regression model known as a *logit model* (there is a huge variety of similar models classified as regression models).

According to Shields and Kures (2007) the model had mixed success. The notable differences that they found showed that closed stores were, on average, in markets that had: (1) fewer households; (2) closer competitors; (3) further from a distributor; and (4) lower poverty rates. They also speculated on variables that might have been included had data been available – factors such as future lease liability and real estate value; store age, size, and capital spending requirements. As with many examples, the models are only as good as the data underpinning them.

A final useful example is provided by Simkin (1989, 1990) who uses measures of competition, trade areas, accessibility, store characteristics and catchment demographics to construct turnover predictions for a variety of retail activities. The model was found to generate correlations (r-squared) of 'not less than 0.81' for major retailers in 'consumer electronics, dry cleaning, fast food and petroleum' (Simkin 1989: note, the closer the r-squared value is to one then the greater is the correlation). Despite their known weaknesses, which are amplified further below, such models are still in development. For example, Bai *et al.* (2009) discuss an application to convenience retailing in China, although in this case success criteria for the model are harder to evaluate.

Although regression models allow greater sophistication and objectivity than more manual analogue techniques there does remain several problems. The primary weakness of such models is that they evaluate sites in isolation, without considering the full impacts of the competition or the company's own global network. As the above bank branch example shows, the level of competition is typically incorporated by the simple absence or presence of stores. A second major weakness is the problem of 'heterogeneity of sample stores'. This was also seen as a problem with analogue techniques. That is, how easy is it to find a sample of stores that have similar trading characteristics and catchment areas? Ghosh and McLafferty (1987) discuss this issue in more detail. A third problem relates to the basic feature of regression analysis which assumes that the explanatory variables in the models (*Xmi*) are independent of each other and uncorrelated. In many retail applications this is not the case – independent variables such as floor space and car parking spaces may be strongly correlated. This can lead to unreliable parameter estimates and severe problems of interpretation. This so-called multi-collinearity problem has received much attention in the literature (see for example, Lord and Lynds 1981; Ghosh and McLafferty 1987). However, through careful analysis and interpretation many of these problems can be overcome. Most poor applications of multiple regression in retail analysis have shown statistical naivety and limited understanding of retail process. Fourth, is the problem of defining the catchment area size, especially in relation to the impact of competitors within a certain distance of a store (the fifth variable in Fenwick's example above). Fifth, and from our point of view the most important limitation, is that regression models fail to handle adequately *spatial interactions or customer flows*. That is, they do not model the processes (spatial interactions) that generate the flows of revenue between residential or workplace areas and retail outlets. Although regression models may sometimes demonstrate impressive descriptive powers (through their ability to reproduce the variation in sales across a network) the absence of any process modelling leaves us sceptical as to their ability to undertake *impact analysis* with any great confidence.

Other limitations of the regression-modelling approach are more technical. These methods are typically much more effective in handling straightforward nominal data, and so where the inputs are either ordinal or categorical[1] then the model is much more difficult to implement. In Chapter 9 we will suggest that *scorecards* provide a much more flexible way to accommodate a full range of potential data inputs.

A number of other multivariate statistical approaches can be applied to the problem of store performance analysis. Space does not allow a thorough review of all these alternative models but the interested reader can look at the core papers cited. For example, Mendes and Themido (2004) discuss the possibilities with discriminant analysis, as well as other classification methods including clustering and decision trees. Artificial neural networks (ANN) is regarded as the most suitable of the available forecasting techniques by Alon *et al.* (2001) (and see also Tchaban *et al.* 1998; Das and Chaudhury 2007; Gao *et al.* 2010 among other examples favouring the adoption of ANN).

However, several qualifications are in order at this point. First, in a limited sense this is not surprising since ANN are typically more highly parametrised, non-linear members of the multivariate analysis stable. In other words, for any given problem with a number of predictors and a dependent variable, ANN will always outperform multiple regression. Second, ANN have been applied with most enthusiasm to sales forecasting problems rather than impact analysis or network planning with which we are mostly concerned here. They are relatively well suited for projecting the ups and downs in a sales trend (for example, under the influence of seasonal variations) but not obviously well adapted to problems with complex interactions. Third, ANN are black box methods and as such generate numbers of varying reliability, but what they typically fail to produce are insights or confidence in a set of conclusions underpinned by sound reasoning.

5.4 Spatial interaction modelling

5.4.1 Background and model definitions

During the 1980s and 1990s, as retail markets became more saturated and the competition for sites increased, there was a significant shift towards the use of more sophisticated techniques in network planning. Improvements in information technology (the increase in power of microcomputers for example), the development of GIS software, increases in quality and quantity of data and the challenges of identifying potential new sites combined to offer a new, favourable environment for spatial modelling techniques. One such modelling technique which became a core feature of many large corporate retail network planning departments was the *spatial interaction model* (SIM). SIMs were originally developed from the gravity-based principles of Newton's scientific theory of universal gravitation (hence they are most commonly called *gravity models* in the retail industry), but the work of Wilson (1970, 1974) showed that these models could be derived using principles from statistical mechanics. These 'entropy-maximising techniques' showed that the final form of the SIM developed by Wilson was based on a sounder theoretical basis, allowing some of the problems of early gravity models to be rectified.

By definition, SIMs are used to simulate or predict the interactions or flows (e.g. people, households, expenditure) between origins and destinations. Wilson (1974) defines four different variations of the SIM in his 'family of SIMs', with each one differentiated by the constraints that are placed on either demand or supply-side variables. Regarding the retail sector, the most commonly used SIM is the production-constrained model, shown below:

$$S_{ij} = A_i O_i W_j exp\left(-\beta d_{ij}\right) \tag{5.3}$$

Where, S_{ij} is the flow of people (or expenditure) from residential area i to retail store j, O_i is a measure of demand (expenditure) in area i, W_j is a measure of attractiveness of retail store j, d_{ij} is a measure of the distance (e.g. time, miles) between i and j, β is the distance decay parameter, and A_i is an internal balancing factor to ensure that all demand is allocated to retail outlets within the region, written as:

$$A_i = \frac{1}{\sum_j W_j exp\left(-\beta d_{ij}\right)} \tag{5.4}$$

The model allocates flows of expenditure between origin and destination zones on the basis of two main hypotheses: (1) Flows between an origin and destination will be proportional to the relative attractiveness of that destination *viz-à-viz* all other competing destinations. (2) Flows between an origin and destination will be proportional to the relative accessibility of that destination *viz-à-viz* all other competing destinations. Thus, the model works on the assumption that in general, when choosing between centres which are equally accessible, shoppers will show a preference for the more attractive centre. When centres are equally attractive, shoppers will show a preference for the more accessible centre. Note, however, that these preferences are not deterministic. Thus when choosing between equally accessible centres, shoppers will not always choose the most attractive. The models are therefore able to represent the stochastic nature of consumer behaviour well.

The cost or distance deterrence term has often been measured simply as straight-line distance in the past. In reality of course, travelling in a straight line is often not possible: there are only so many places where it is possible to cross the River Thames in London, for example. If there is no bridge to connect them, then two places 100 m apart might actually be five or six miles apart by the road network. Today, the availability of digital road networks has enabled some organisations to produce travel time matrices, based on average speeds attached to these networks. Given that these travel times are likely to take account of barriers such as rivers, railway lines, motorways etc., they should be used in preference to straight-line distance when available (see again the discussion on GIS and network analysis in Chapter 3).

Some applications of SIMs produce very complicated cost terms. In some of the shopping models produced by the UK Unit of Retail Planning and Information in the 1970s and 1980s (for local authority use in the main), the cost term was often disaggregated by person type. For car owners, the function was a reasonably straightforward combination of road distance and travel time. For public transport, however, the function included 'in vehicle time', plus 'walking time', plus 'waiting time', plus actual fares. The difficulty and expense of collecting such data would put most people off using these elaborate functions. It is also far from proven in the literature as to whether this level of complexity significantly increases model accuracy.

5.4.2 Model calibration

The validation, or calibration, of a SIM is an important process to assess the ability of the model to replicate known data sets. Calibration uses statistical methods to find parameter values for the model so that the interaction set that it produces is as close as possible to known interaction patterns. In the model above (equations 5.3 and 5.4) the main parameter of the model is β, which controls the importance of distance in the model.

Different patterns of customer flows may be represented within the model using variations in the distance deterrence parameter (β in equation 5.3). The β parameter measures the willingness or ability of consumers to travel over the landscape. This can allow markets with very different characteristics to be modelled within the same framework; for example, Figure 5.1 compares the catchment area within Leeds for two products – books and newspapers. The interaction pattern for books would be represented with a much lower value for β than for newspapers because people are generally more willing to travel further to buy books.

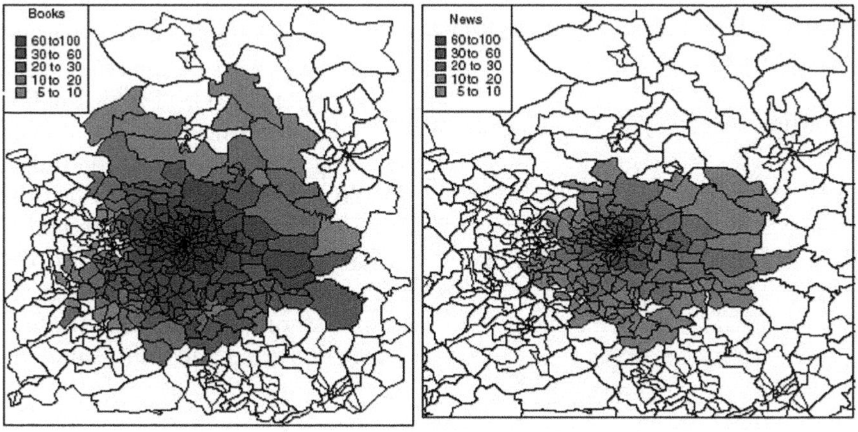

Figure 5.1 Catchment areas for two different products (books and newspapers) in Leeds, UK.

Source: Authors

If the model is well calibrated, the pattern of customer flows from individual demand areas to individual supply locations will be faithfully reproduced. Figure 5.2 gives an example of how it is possible using disaggregated SIMs (see Section 5.4.5) to reproduce known flows based on a retailer's actual data or perhaps survey data (Figure 5.2(a)) with estimated flows (Figure 5.2(b)) – these are based on Newing (2013) and the study of retail interaction patterns in Cornwall, UK (Newquay being the site of the store modelled in this example). If the individual flows are being modelled accurately, when they are summed at each supply point to provide an estimate of store revenue, this should also be expected to be accurate. This is obviously vital for retailers, who will wish to have maximum confidence in any revenue estimates to be provided by the model for new store openings, or indeed refurbishments, relocations or closures.

Birkin, Clarke and Clarke (2002) showed, in a modelling application for a DIY retailer in the south of England, that predictive forecasting accuracy on store sales of ±8 per cent could be achieved. Clarke and Clarke (2001) report on the results from a similar exercise applied to modelling fuel volume for petrol stations in Sicily. This is an extremely complex task, given the nature of fuel purchasing. Whereas supermarkets and other large retail outlets are largely dependent on a local catchment population whose extent and purchasing power can be systematically evaluated, there are far more unknowns with fuel retailing. Nevertheless, Clarke and Clarke (2001) were able to demonstrate modelling accuracy of ±14 per cent for this more complicated application.

So how does the calibration process work? Normally, the user would need to compare real world flows – S_{ij} (obs) – with those produced by the model – S_{ij} (pred). Observed data might come from a number of sources such as store loyalty cards or survey data. One common statistical procedure to calibrate the models is the 'sum of squares' routine. The idea here is to find a value of β which will minimise the difference between observed and predicted flows: or mathematically,

$$\text{Minimise } S = \Sigma_{ij} \, [S_{ij}(\text{obs}) - S_{ij}(\text{pred})]^2 \tag{5.5}$$

To determine how close the predicted and actual values, many other goodness-of-fit statistics have been used in the literature within SIM (see Batty 1976; Knudsen and Fotheringham 1986 for good reviews). R^2 is one of the most commonly used goodness-of-fit statistics and is formulated as follows:

$$R^2 = \left[\frac{\sum_i \sum_j \left(S_{ij} - \bar{S}_o\right)\left(\hat{S}_{ij} - \bar{S}_m\right)}{\left[\sum_i \sum_j \left(S_{ij} - \bar{S}_o\right)^2 * \sum_i \sum_j \left(\hat{S}_{ij} - \bar{S}_m\right)^2\right]^{1/2}} \right]^2 \tag{5.6}$$

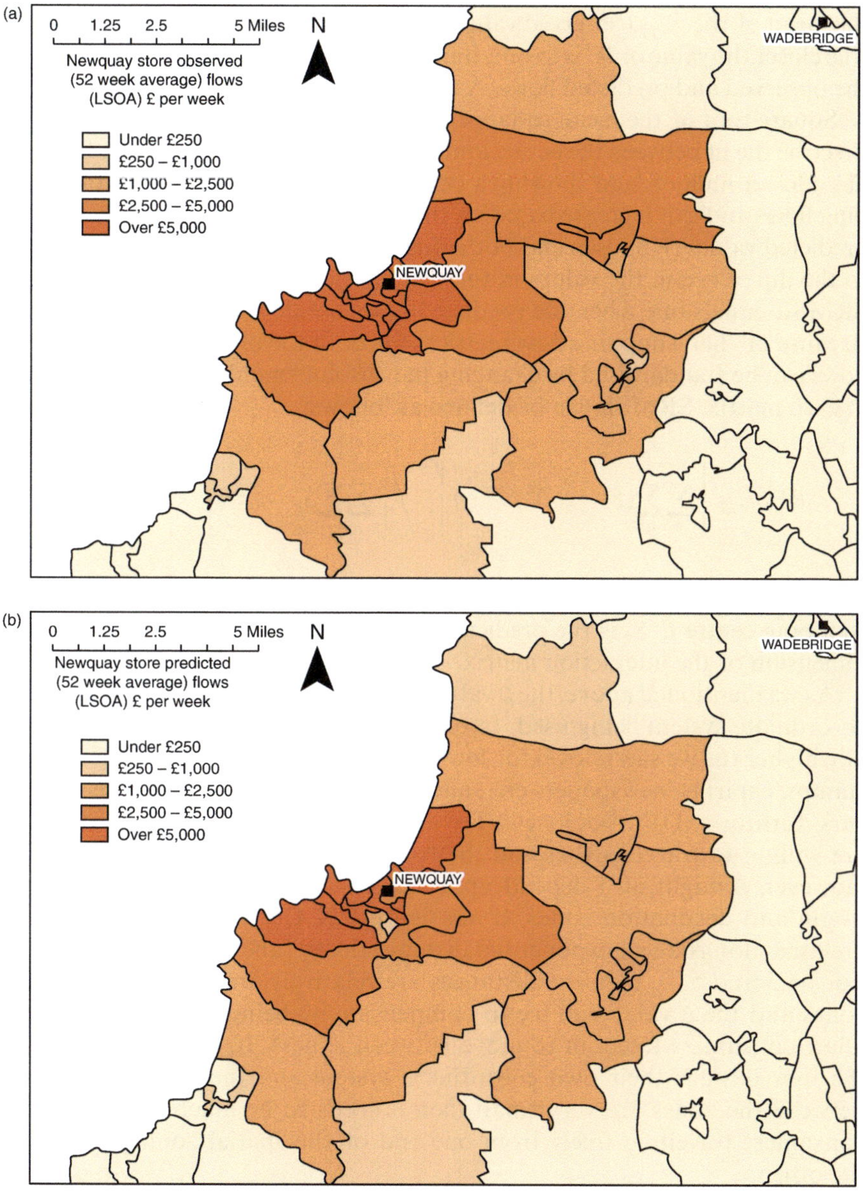

Figure 5.2 Comparing retail flows in Newquay, Cornwall: (a) observed and (b) predicted by a SIM.

Source: Newing (2013)

Where, $\bar{S}_o$ represents the mean of the S_{ij} s (predicted) and $\bar{S}_m$ represents the mean of the S_{ij} s (observed values). R^2 values range between zero and one. The closer the value of R^2 is to one, the more exact the correspondence between the observed and predicted flows. A zero value reflects no correspondence.

Square root of the mean square error (SRMSE) is also commonly used to describe the fit between the actual and predicted interaction matrices. SRMSE has a lower limit of zero showing a perfectly accurate prediction of the shopping flows and can increase based on the difference between the observed and predicted values (Knudsen and Fotheringham 1986). SRMSE is very sensitive to the difference in the values in the two matrices; for instance, the value of SRMSE equals one when the predicted flows are double the observed flows. Because of that, shopping flow matrices for both the observed and predicted flows can be standardised by changing the absolute numbers into percentages in each matrix. SRMSE can be defined as follows:

$$SRMSE = \left[\sum_i \sum_j (S_{ij} - \hat{S}_{ij})^2 / m\,x\,n\right]^{1/2} \Bigg/ \left[\sum_i \sum_j S_{ij} / m\,x\,n\right] \tag{5.7}$$

Where, S_{ij} represents the actual shopping flows from postal sector i to shopping centre j, S_{ij} is the predicted flow between i and j, and $m\ x\ n$ is the dimension of the interaction matrix.

As we mentioned above, the β value depends upon the application and the co-ordinate system being used. In terms of the application, β will be generally higher (as we saw above) for lower order goods, such as milk, groceries, stamps, aspirins, newspapers etc., and lower for higher order goods such as cars, furniture, DIY, books etc. For the latter range of goods, consumers are willing to travel further and the friction of distance is less important. However, β might also depend upon the co-ordinate system used for the origin and destination zones. If six figure UK OS co-ordinates/ordinates are used alongside an exponential distance decay function (as seen in equations 5.3 and 5.4), then the distances are relatively large between any two zones and the β value will try to compensate by being generally lower (to allow consumers freedom to move between zones). If, on the other hand, the user designs their own co-ordinate system and the relative distances between the zones are very small then β needs to be much higher to stop consumers travelling freely from one end of the spatial zoning system to the other.

The choice of an appropriate β value creates a problem when using 'black box' SIMs without any calibration procedures. This is the case with GIS packages such as Arc. Although it is good to see GIS vendors trying to improve the level of analysis in GIS packages, by adding such models, there are many dangers in offering a standard (aggregate) model to potential users. Clearly the user cannot disaggregate or change the model in any way. More worryingly from a calibration point of view is when vendors suggest a value for β – say 2.00. The real value could, in fact, be anything!

It is possible, and feasible when data provision is good, to disaggregate β by origin zone or perhaps person type (car owner/non-car owner for example). For the former, this means that for each origin zone there is a unique β value. This reflects the fact that consumers from some zones are more likely to make shorter or longer trips than consumers from other zones. The most obvious example is the difference between rural and urban areas. In urban areas there are generally more goods and services available forming a larger choice set and residents in urban areas do not travel long distances for retail goods (or do not have to in most cases). On the other hand, rural residents may be forced to travel long distances to the nearest superstore or DIY outlet. For them, the choice set is limited and their willingness to travel long distances is greater – hence a different distance β value. Birkin, Clarke and Clarke (2002) show an example of a model built for the car market where β is disaggregated by origin and destination zones, person type and model segment.

5.4.3 Model outputs

So how would a retailer typically use such models? First, as discussed in Section 5.4.2, the model is calibrated to reproduce existing interaction patterns between populations in residential zones i and stores or shopping centres j. Thus, the retailer can examine the catchment areas of existing stores and estimate market shares and expected revenues (and potential new ones as we shall discuss below). Figure 5.3 shows the typical interaction patterns around a particular supermarket in the small coastal town of Looe in Cornwall, UK. The map shows the market share of a Morrison's store across the study area estimated by the SIM. The key advantage of the SIM is that the model takes into account the location and nature of the competition (i.e. all stores present in the region are modelled) and thus overcomes many of the problems of GIS catchment areas produced using buffer and overlay techniques (see Section 3.3). Figure 5.3 shows that this results in more realistically shaped catchment areas (not circular ones). In this case trade for the centre quickly falls off to the north and east as competition from other stores in Looe and Liskard becomes fiercer.

Once the flows have been estimated then the SIM can produce revenue estimates for every store or centre in the system. The revenue of a particular store is the sum of the flows coming into that centre from all origin zones. It can be expressed mathematically as:

$$Dj = \sum_i S_{ij} \tag{5.8}$$

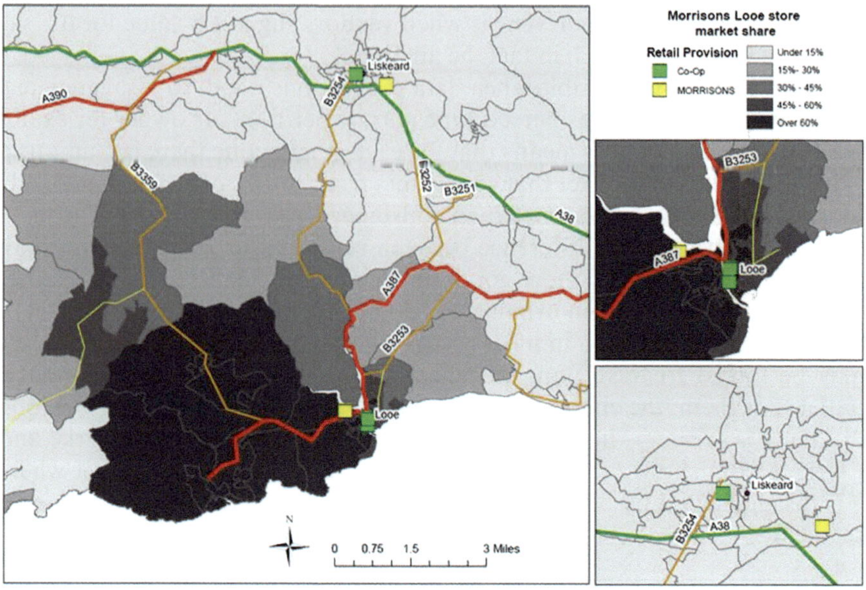

Figure 5.3 Using SIMs to estimate small-area market shares.
Source: Newing (2013)

All store revenues or turnovers may not be available from published sources: few retailers have turnover estimates for their competitors for example. Thus the model estimates gives the retailer valuable insights into how well or badly their competitors are likely to be performing, store by store. It will also help them to benchmark the performance of their own stores. That is, given a certain population size and type, and the nature of the distribution of all competitor outlets, what would the model expect a certain outlet to be achieving in sales terms? This helps to provide a more objective picture of store potential. Is a store which turns over $3 million per annum doing well or badly in relative rather than absolute terms? Table 5.1 shows an example of model results versus actual sales for a retailer in the county of Essex, south-east England. Once the model has been calibrated as thoroughly as possible, residual values between observed and predicted can reflect over and under-performance. Thus, the retailer might ask why their store in Walthamstow is doing so well compared to model predictions. Is the store simply very well managed or are there more local factors causing higher than expected performance (a bus station nearby or some other factor causing a high footprint)?

The models are, however, most often used in what-if? fashion. Having identified the variations in market share, the retailer may be keen to improve its performance by opening new outlets in the areas which currently have a

Table 5.1 Using SIMs to benchmark sales for retailer X in Essex, UK.

Location	*Predicted revenue*	*Actual revenue*	*Store performance*
New Southgate	£4.4m	£4.5m	102
Walthamstow	£5.7m	£6.8m	118
Hendon	£3.7m	£3.6m	97
Harlow	£3.1m	£2.9m	95
Colchester	£3.1m	£3.0m	98
Basildon	£2.2m	£2.3m	103
Chelmsford	£1.9m	£1.8m	97

Source: Authors

Table 5.2 Modelled (estimated) impact of a new Tesco store in Looe, including impacts on existing stores in the region.

25,000 sq ft Tesco on Barbican road out-of-town site, east Looe			
Revenue and trading at new store (Tesco)		*Impact on Co-op*	
52 week average	£351,077	Sales (Looe)	Fell by 63.2%
Sales/sq ft 52 week average	£15.01 (£15.91)*	Market share (Looe catchment)	Fell by 65.6%
Sales/sq ft January	£11.09	Market share (countywide)	Dropped 0.5% to 10.5%
Sales/sq ft August	£19.74	*Impact on Morrisons*	
Average trip distance	5.01 km (6.25 km)*	Sales (Liskeard)	Fell by 18.1%
Looe catchment market share	61.3%	Market share (Looe catchment)	Fell by 60.1%
Tesco market share countywide	31.9%	Market share (countywide)	Dropped 0.7% to 18.7%

Note: * Values in brackets represent modelled company average for Cornish Stores

Source: Newing (2013)

low market share, or maybe interested in modelling the impacts of a competitor store opening or closing. The models can then be used to test the impact of a new store opening in a more comprehensive fashion. Returning to our Looe example above, Table 5.2 shows the estimated results of a new Tesco superstore opening in the town, and the impacts on the two existing supermarkets present.

The resulting loss in market share for Co-op or Morrison's can also be mapped. Figure 5.4 shows the estimated loss of market share for Morrison's following the Tesco opening. This would effectively wipe out Morrison's sales in the eastern part of the region.

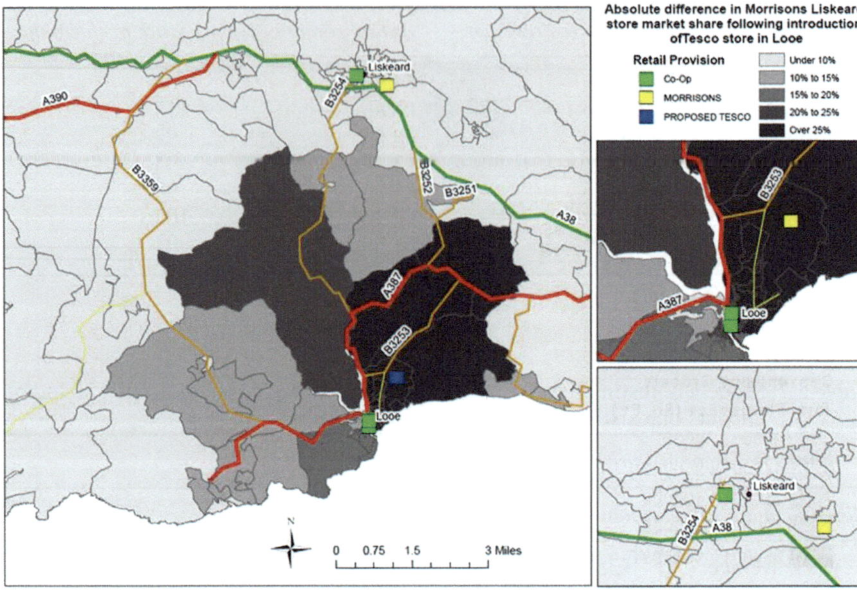

Figure 5.4 Estimated loss of market share for Morrison's following the new Tesco store opening.

Source: Newing (2013)

5.4.4 SIMs used for accessibility analysis

Finally in relation to outputs, the SIMs can be specified as interesting residence-based or supply-side-based performance indicators (Bertuglia *et al.* 1994; Clarke and Wilson 1994). On the residence side we can use them as provision indicators, examining how well served residents are in terms of access to retail opportunities. The aggregate level of provision in an area can be given as:

$$w_i^m = \sum_j \frac{S_{ij}^m}{S_{*j}^m} W_j \tag{5.9}$$

Level of provision per household is an indicator that divides the aggregate level of provision by the number of households in the residence zone, as follows:

$$v_i^m = \frac{w_i^m}{H_i^m} \tag{5.10}$$

A good example of the use of these indicators has been on work relating to 'food deserts' – areas with poor access to main grocery stores where prices

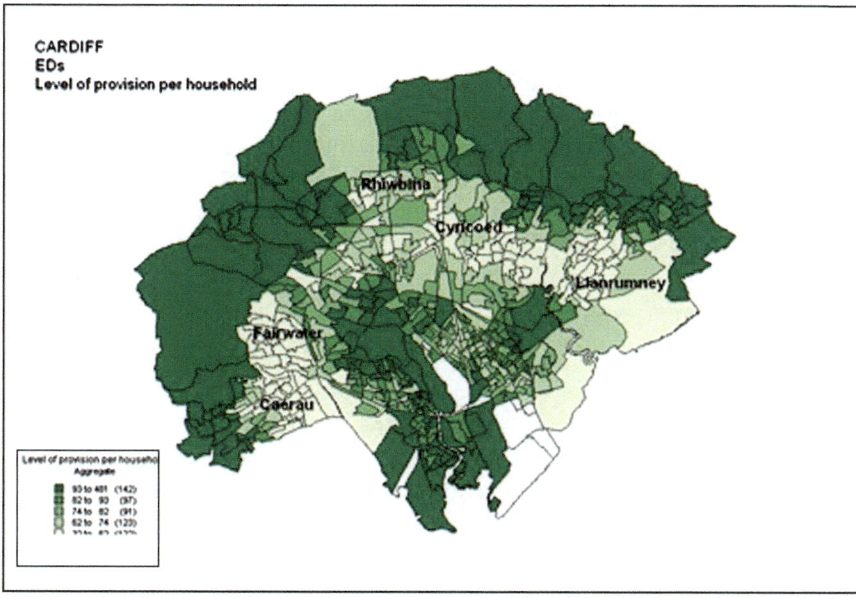

Figure 5.5 Estimating consumer access to grocery stores in Cardiff, UK.
Source: Clarke *et al.* (2002)

may be higher and the choice of fresh fruit and vegetables poorer (thus raising issues relating to poor diet and obesity), if provision is largely in the hands of small independent retailers. The debate on whether such areas exist or not is ongoing (see a range of opinion in Wrigley 2002; G. Clarke and Bennison 2004; Cummins *et al.* 2005; Shaw 2006; Larsen and Gilliland 2008 for example). However, we would argue that these interaction-based indictors are a useful tool for measuring poor access. Alternative measures such as provision or floor space per head of population are too simple especially when there are many small zones with zero scores. Figure 5.5 shows the outcome of calculating the provision indicator in Cardiff, Wales (G. Clarke *et al.* 2002). The patterns of access and provision are clearly highest at the suburban edge of the city where the majority of corporate food stores have been located (although there are gaps in the northern outer suburbs where council houses were built after the war for lower income residents, areas avoided by the same corporate players).

5.4.5 Model disaggregation

Over time, a number of derivations of the SIM have been developed to ensure the technique is applicable in different sectors of the retail market. It has become common place for modern SIMs to be highly disaggregated, to account for more complex behaviour in a given system (in addition to disaggregation of the β term discussed above). Wilson (1974) shows how to more accurately represent different behaviour in commuting to work or shopping, for example, through the disaggregation of the interaction model to represent different modes of travel in a transport model. Since then, SIMs have been disaggregated or modified in accordance with different retail sectors and channels. In the rest of the book we will introduce new ways that models, or the main variables within them, can be disaggregated. For now it is useful to specify a more fully disaggregated model: modifying equation (5.3) we now have

$$S_{ij}^{m} = O_{i}^{m} A_{i}^{m} W_{j}^{\alpha^{m}} \exp(-\beta^{m} d_{ij}) \tag{5.11}$$

where:

S_{ij}^{m} = expenditure by household type m in residence zone i at destination j
O_{i}^{m} = level of consumer expenditure of household type m in residence zone i
A_{i}^{m} = a balancing factor to ensure that:

$$\sum_{j} S_{ij}^{m} = O_{i}^{m}$$

which is calculated as:

$$A_{i}^{m} = \frac{1}{\sum_{j} W_{j}^{\alpha^{m}} \exp(-\beta^{m} d_{ij})} \tag{5.12}$$

W_{j} = the attractiveness of destination j
α^{m} = a parameter reflecting the perception of a destinations attractiveness by household type m
d_{ij} = the distance between origin i and destination j
β^{m} = the distance decay parameter for household type m

This model suggests that different household types (i.e. age groups or social class groups) will have different retail behaviour patterns and that these can be captured by the models. More discussion of this type of model disaggregation is given in Birkin *et al.* (2010) and Newing *et al.* (2015). It is interesting to report that recent research on SIMs includes attempts to incorporate parameters on both individual origin zones and destination zones. These models draw upon recent work on geographically weighted regression models (where the key parameters are allowed to vary spatially) and aim to build in many more parameters that can be calibrated when the data is good (Nakaya 2001; Fotheringham 2013; Kordi 2013).

5.5 Other behavioural models

In the 1970s a number of spatial modellers began to build critiques of SIMs because of their lack of behavioural content (despite the ability to disaggregate them as shown above). Oppewal and Timmermans (2001) note that progress with categorical data analysis offered the opportunity to combine economic and psychological theories of consumption and individual choice behaviour through a series of individual *discrete choice models* (see also Ben-Akiva and Lerman 1985; Wrigley 1985). Discrete choice models aim to predict the probability that an individual will choose a particular retail destination on the basis of the characteristics of that person (age, car owner/non-car owner etc.) and the attributes of the store (size, car parking etc.). They are thus similar to SIMs but built from the bottom-up (i.e. from the level of the individual). They are also very data hungry – it is necessary to build large-scale surveys of individual shopping behaviour for any meaningful study. A key problem is also how to aggregate? From a survey of 2,500 people one could have 2,500 discrete choice models – thus careful thought is required on how to construct a robust set of models – that is, a set of models for different age groups, for car owners vs non-car owners etc. It might also be noted that when SIMs are highly disaggregated (as we shall see more and more in the rest of the book) there can be very little difference in practice to the structure of these discrete choice models. Harry Timmerman's retail research group at the University of Eindhoven in the Netherlands are major users of discrete choice models in retail analysis and the interested reader should look at his website for a large number of publications.

5.6 Conclusions

It has been argued that SIMs are more powerful predictors of store turnovers and spatial interactions than any of the other methods introduced so far. However, despite the considerable benefits associated with SIMs, care must still be taken when using them. For instance, SIMs are data intensive, which means they require good quality data (demand and supply). Poor quality data causes issues with calibration and ultimately accuracy. Even if it is possible to gain access to company loyalty card data, calibrating a model exclusively to one retailer's customers rather than the whole market could create a bias which may manifest itself in a false value of a parameter for other competitor types (Birkin, Clarke and Clarke 2010). We have also noted the importance of disaggregating SIMs to fit the case study under investigation. Given the complexities of consumer behaviour for different types of goods and services it is unlikely that a model which fits the car market, for example, can be taken off the shelf to operate in the cinema market. Thus the user needs to be a skilful spatial modeller or outsource to those that are. More examples of model disaggregation appear in the rest of the book.

So do retailers use SIMs or gravity models? The recent work of Reynolds and Wood (2010) and Wood and Reynolds (2011; 2012) is again useful here (see Chapter 1). The results of their 2010 survey show that, of the retailers surveyed, 80 per cent use analogues, while 65 per cent use statistical approaches and 65 per cent again use SIMs or gravity models. So the number seems to have grown a little since the Hernadez and Bennison survey of 2000. In addition, we should note that many retail consultancies also offer such modelling capabilities. Thus they are an important part of the kitbag of techniques in UK companies such as CACI, GMAP, Experian, JAVELIN, Polk etc. While these modelling methods will never replace the use of experience or gut feeling, it is encouraging to see more engagement with the methodologies regularly used by academic geographers.

Note

1 Nominal data is both ranked and scaled – if I have an area with unemployment of 20 per cent then this is twice the level of an area with 10 per cent, and four times an area with 5 per cent; and so on. Ordinal data is ranked, but not scaled. If you ask people how much they like shopping at Marks and Spencer on a scale from 1 (don't really like it) to 5 (absolutely love it) then customers scoring 5 will probably shop more often at M&S than those scoring 1; but whether they are five times as likely, or twice as likely, or just a little more likely, we simply can't say. Categorical data is just different. For example, if we were to distinguish stores according to whether they are in a high street, part of a covered centre, in a small cluster or isolated, then we can't say much a priori about these groups, but we have an expectation that the type of location might affect their performance.

6 Exploring retail demand

Estimation methods and future drivers of change

6.1 Introduction

In the preceding methodology chapters (3–5) we have seen that retail location planning should involve a consideration of demand and supply-side factors. These are crucial for any form of decision making, whichever form of retail location technique is chosen. The aim of this chapter is look in more depth into how we might measure retail demand within existing or potential new store catchment areas. There are few retail products on the market where demand is ubiquitous – that is, everyone purchases the same amount each day or week. If that was the case, just taking the population itself would be the simplest proxy for demand. For most goods and services demand needs to be disaggregated, and indeed most retailers actually target different population groups. Table 6.1 shows an example of niche marketing undertaken by the then UK Burton high street clothing group in the mid-1980s – now part of the Acadia group: first shown in Birkin *et al.* (2002). Although somewhat dated, it still shows how a major corporate chain differentiates the market, and hence its fascias, by age and income type (where 'sophisticated' in this instance means more expensive). Thus, demand needs to be considered very carefully by the retail location analyst, especially in relation also to where that demand might reside. The task for Burton was to find the right high street for each fascia (or outlet in Table 6.1), or find the right fascia/outlet for each high street.

In this chapter we explore the concept of retail demand in more detail, examining issues relating to how we might estimate small-area demand and how that demand might itself change in the future. The chapter uses the spatial interaction model (SIM) as a framework for discussion (in terms of how different measurements of demand can be handled in store location planning) but the issues are appropriate for any form of retail assessment method.

In the aggregate SIM framework (see Chapter 5) demand looks deceptively simple: O_i, where this represents the total number of persons or money available to spend on retailing in zone i. As we shall see, this term can be defined in many ways and can be disaggregated according to the type of application. We shall explore these issues below.

Table 6.1 An example of niche marketing: matching person types to retail fascia.

Outlet	*Market*	*Profile*	*Target market age*	*Total/ no. outlets*
Burtons	Men Youths	Mainstream formal and casual clothing and accessories	20–40	467
Alias	Men	Fashionable colour co-ordinated leisure and casual clothing	25–40	33
Top Man	Men Youths	Fashion-aware formal and casual clothing and accessories	11–30	252
Radius	Men	Individualistic formal and casual clothing and accessories	20–35	32
Principles for Men	Men	Modem classic clothing and accessories for business and pleasure	25–45	105
Champion Sports	Men Women Children	Fashionable sportswear and equipment	15–35	100
Dorothy Perkins	Women	Mainstream formal and casual clothing and accessories	18–40	467
Secrets	Women	Fashionable lingerie and nightwear	15–45	27
Top Shop	Women	Fashion-aware casual and occasion clothing and accessories	11–30	302
Principles	Women	Sophisticated fashionable classic clothing and accessories	25–45	192
Evans	Women	Mainstream fashion size 14+	25–60	196
Debenhams	Men Women Children Home	Mainstream mass-market fashion for the individual and home	All ages	71
Harvey Nichols	Men Women Children Home	Exclusive and designer fashion for the individual and home	All ages	1

Source: Burton (1988) reproduced in Birkin *et al.* (2002)

6.2 'Direct' demand data

In some markets, such as automotive distribution, the requirement for customer registration means that existing patterns of demand for a product may be used as a proxy for future patterns. In the car market, for example, there is a legal requirement in most countries of the world for owners to register/tax their vehicles. Thus the government has a very good data base on car

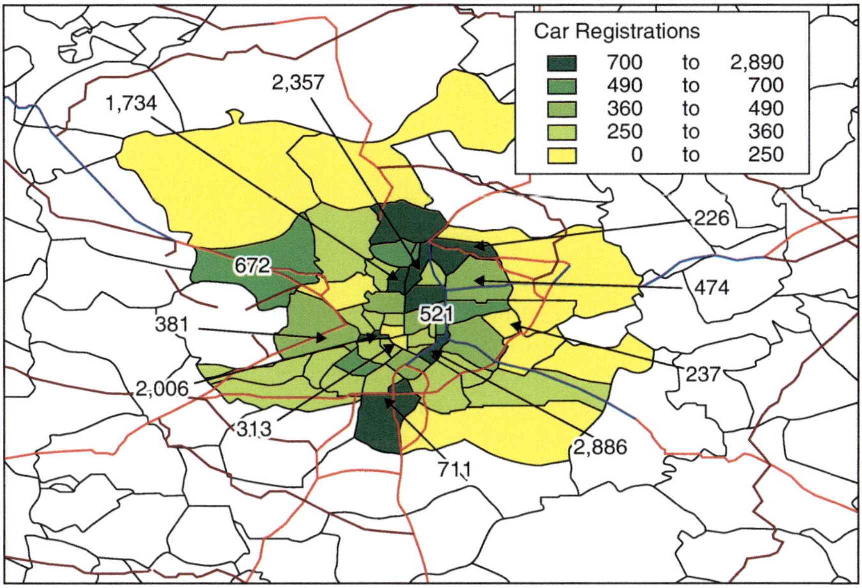

Figure 6.1 New car registrations in Madrid by postal area over a 12-month period.
Source: Authors

ownership. In the UK, and indeed many other countries, the government licensing authorities provides or sells this data to the vehicle manufacturers. This provides a unique retail data base – the sale of every new car by dealer and postal address, for a particular manufacturer. Thus, it is a commercial market tailor-made for GIS and spatial analysis (Birkin *et al.* 1996). Figure 6.1 shows the total new car market for Madrid in Spain over one calendar year. By monitoring the total market over time, analysts can build up a picture of year-on-year demand for new cars by postal area – thus the demand estimates for the following year can be thought of some kind of average over the last three, five or ten years. There is no need to do any further simulations as in most other markets, except factor in lower sales in times of economic recession.

6.3 Estimating demand

In other markets, where such data are not routinely collected (for all customers), demand estimates must be constructed by combining information about acquisition propensities, aggregate market sizes and small-area population data. The first approach is to combine raw census data with market survey data. Thus, for example, we might know from a survey such as the UK 'Family Spending' survey that the average household spends £60 per week on groceries. If there are 2,000 households in a zone then the total expenditure available for grocery spending is £120,000.

However, it is usually possible to do better than that. As noted above, in many cases demand needs to be disaggregated by person or household type. This can be done in a number of ways. First demand can be disaggregated by gender. Clearly retailers aim certain products at different gender groups (men's versus women's clothing for example). Products may even be differentiated within seemingly ubiquitous products such as cars. For example, in the UK over 60% of Minis and Fiat 500s are bought by women (Telegraph 2013).

Demand can also be disaggregated by age. There is an interesting literature now on the topic of targeting different age groups (for example Moschis *et al.* 1997; Taylor and Cosenza 2002; Meneely *et al.* 2009). We shall explore this in more detail below in relation to the elderly population. Again certain products are more obviously age-related than others. Toys R Us stores, which can now be found all over the world, clearly target young families with two or three children per household.

Perhaps the most common disaggregation of demand is by social class or by income if that data is available. Again some retailers are known to target higher income professional or managerial consumers (Sainsbury's, Waitrose and Marks and Spencer are obvious examples in the UK grocery market) while the discount market continues to grow, initially from targeting the lower income groups (e.g. discounters such as Aldi and Lidl in the UK). Similarly, Table 6.1 showed that Burton differentiated the market using the term 'sophisticated'; a reference to higher price in particular. The UK 'Family Spending' survey again can be matched with the Census of Population to produce estimates of spending by social class. The following data (£ per week) can be used, for example, to estimate expenditure values based on the number of households falling into each social class category (where AB=professional/managerial and DE=unskilled workers)

AB – £75; C1 – £64; C2 – £58; D – £55; E – £51

Suppose we have two zones with the same number of households (750). Disaggregating by social class will differentiate between high and low spend given the same number of households: that is, in zone A we may have the following distribution of households:

AB=200, C1=150; C2=250: D=100; E=50

Then the total weekly expenditure available in this zone is

(200*75)+(150*64)+(250*58)+(100*55)+(50*51) = £47,150

and the mean value is £62.86; In zone B we may have

AB=50, C1=100; C2=250: D=150; E=200

Then the total expenditure available in this zone is

(50*75)+(100*64)+(250*58)+(150*55)+(200*51) = £43,100

and the mean value is lower at £57.4 per week

Another important factor in retail spend may be ethnicity. In most city regions of the world there is a growing ethnic mix within different residential areas. Figure 6.2 shows the ethnic distribution patterns for census wards in London (Stillwell 2010). Clearly retailers need to understand the ethnic market. Some ethnic groups will be unlikely to purchase certain products while providing a real niche market for others.

It is interesting that as countries have opened up borders to immigrants (especially in Europe) there has been a new wave of ethnic retailing over the last decade. Wrigley *et al.* (2009) and Wrigley and Dolega (2011) have documented the growth of independent convenience outlets in certain UK high streets, which seems to be at odds with the general perception that independently owned outlets have been drastically reduced by the tactics of the large multiple retailers, both opening large out-of-town outlets and, more recently, entering the high street convenience market (see Chapter 2

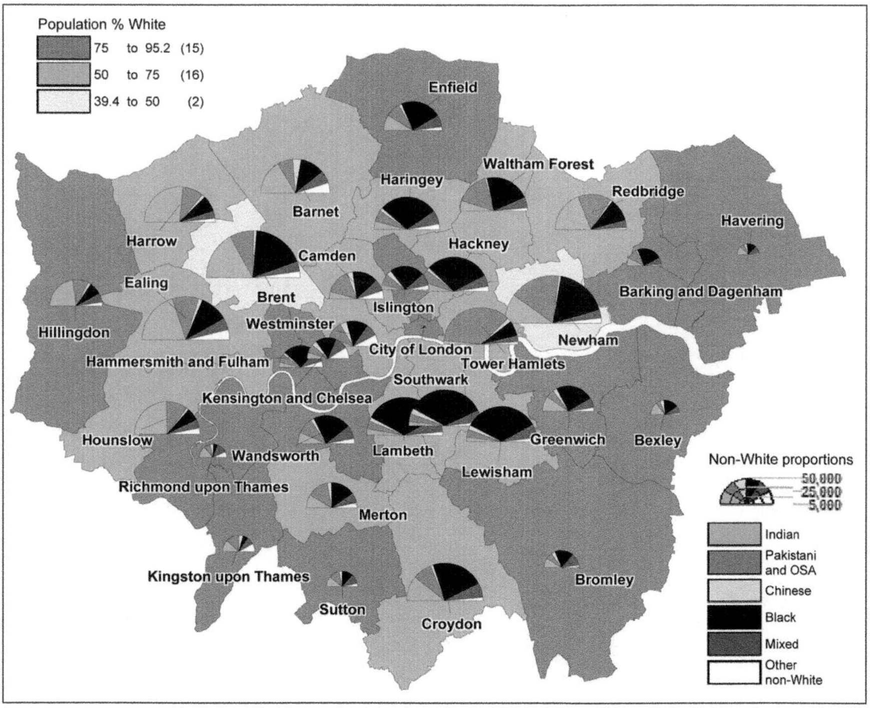

Figure 6.2 Ethnic distribution in London.

Source: Figure created by John Stillwell. (OSA = Other South Asian) and Hood *et al.* 2016). The growth of a new wave of ethnic retailers in the UK (largely supporting new communities from Eastern Europe) goes a long way to explain this phenomenon (Figure 6.3; see also Guy 2008; Hall 2011; Wrigley *et al.* 2017).

This method of estimating demand (combining census and survey data) is simple, robust and reasonably quick to carry out. However, by categorising individuals or households by age, social class etc. we are in danger of ignoring the types of areas they live in, that is, ignoring for instance that manual workers in an inner-city area of Leeds could reasonably be assumed to exercise a different level of spending power than their counterparts in a more affluent suburb. There are additional concerns. First, in the past, social class (as defined by the census) has often been used in market research as a proxy for affluence. Increasing social mobility in recent years makes such proxies increasingly unreliable. Second, an inherent difficulty exists of recognising the regulatory processes that determine variations of affluence over space (e.g. land and house prices). The use of geodemographics and/or direct income estimates might thus be appropriate to undertake (see Chapter 4 and Sections 6.4 and 6.5).

Figure 6.3 Ethnic growth providing new retail opportunities on the UK high street.
Source: Unattributed photographer

Another issue that might impact on these types of demand estimates are unreliable surveys. The size of a survey is important, as too is the degree of reliability associated with consumer response. This brings to mind the 'stated versus revealed preference' issue. For example, in relation to alcohol it is often the case that the average man can be expected to exaggerate the amount he drinks (unless filling out health forms!), while a woman normally underestimates alcohol consumption. Such patterns occur by degrees. Part of the demand estimation exercise must always be to assess where the true dividing line between stated and revealed preference lies.

6.4 Geodemographics for demand estimation

As noted in Chapter 4, geodemographics has been widely used in retail location analysis and marketing. Although we touched on the use of geodemographics for demand estimation in Chapter 4 it is useful to develop this argument in this section. Suppose, for example, we wish to locate a new discount retail store in the Yorkshire and Humber region of the UK (note the similar example using GIS in Chapter 3). We could estimate demand for discount retailing based on age and social class as described above. However, geodemographics gives a ready-made profile of the population by age, social class and many other dimensions. Thus for discount retailers we might use a classification such as 'constrained by circumstances' as a good proxy for the demand we are looking for. Figure 6.4 shows Figure 2.3 again and maps the distribution of this population group against the existing location of food discount retailers in Yorkshire and Humberside and in London. Armed with this information, the retailer could look for gaps – areas of dark shading with few competitors.

6.5 Exploiting new techniques: microsimulation

Quantitative geography is constantly evolving and new techniques come along from time to time which might offer new solutions to old problems. One such technique is *spatial microsimulation*. Microsimulation is a technique well suited for estimating 'missing' data, which of course is often the case with retail demand data (or income data if unavailable from a census).

Microsimulation can be defined as a methodology that is concerned with the creation of large-scale simulated population microdata sets for the analysis of policy impacts at the individual or household level. Spatial microsimulation techniques involve the merging of small-area census data with national individual or household survey data to simulate a population of individuals within households (for different geographical units), whose characteristics are as close to the real population as it is possible to estimate (G. Clarke 1996; Williamson *et al.* 1998; Ballas and Clarke 2001; Ballas *et al.* 2005, 2007). The survey data provide the crucial interdependencies between household attributes not seen in the census data. When reweighted according to the census

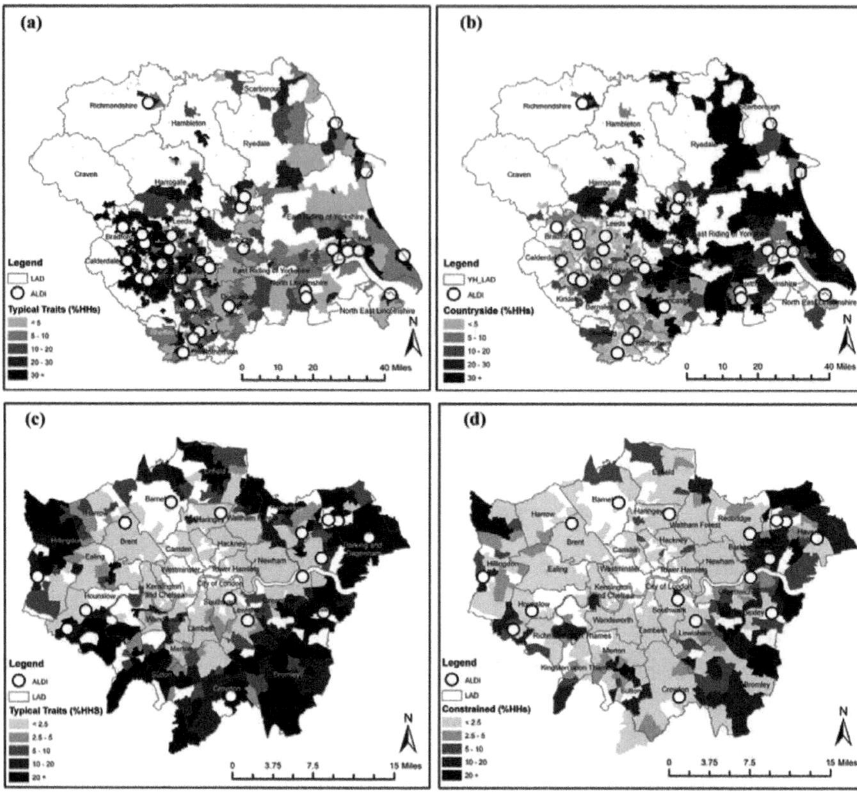

Figure 6.4 Mapping low-income consumers using geodemographics.
Source: Thompson *et al.* (2012)

data, they provide a powerful information set for each household, including many previously 'missing' data components (tax, income etc.).

There are many methodologies for building microsimulation models. The most common methodology today is population reweighting. This involves taking major national surveys and effectively matching those survey households to small-area census data based on common variables (age, sex, social class etc.). Thus every household in a region can be given extra attributes (beyond the main census style variables) based on its most similar household from the survey. The larger the survey, the more accurate the spatial matching.

One good illustration of the strength of this technique for retail demand estimation comes from Nakaya *et al.* (2007). They had access to a large-scale retail survey of customer location preferences for different supermarkets in Kyoto, Japan. Table 6.2 shows the type of household structure within the survey. If the households in the survey could be matched to households in

Table 6.2 Lifestyle groupings in the Kyoto survey.

Younger single	Single	18–39
Older single	Single	>40
Younger couple	Two	18–39
Older couple	Two	>40
Younger family	Two +	18–39
Older family	Two +	>40

Source: Nakaya *et al.* (2007)

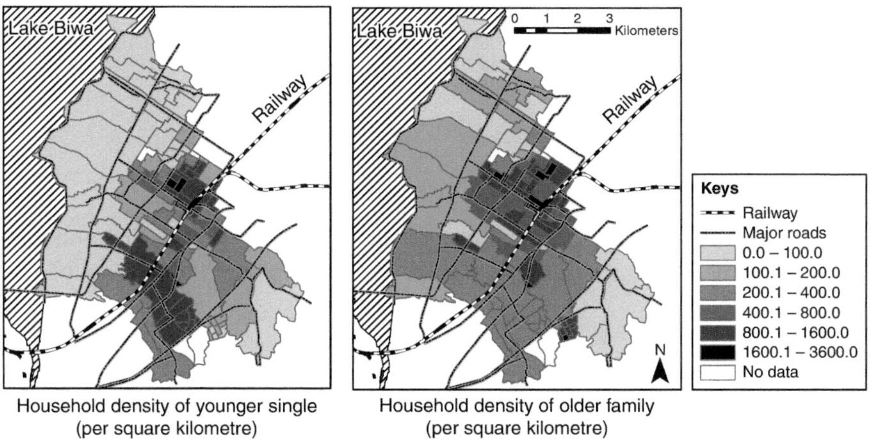

Figure 6.5 Estimating the location of the survey lifestyle groups across Kyoto using microsimulation.

Source: Nakaya *et al.* (2007)

the census (100 per cent coverage) then the authors could estimate shopping behaviour for every household in the city (a powerful tool for retailers and planners alike). Unfortunately, the census household typology did not match exactly. Therefore the authors used microsimulation to estimate the distribution of those households by, in effect, creating new census tables from existing two-dimensional tables, thus matching on a variety of new variables simultaneously (i.e. age, sex, household size).

The result of this simulation is an estimate of the distribution of these different lifestyle groups across and throughout the city. Figure 6.5 shows the patterns estimated for two of these lifestyle groups.

Associated with the estimation of retail demand, there are a number of microsimulation papers that have attempted to model household income, mindful of the inadequacies of both commercially available income data and available census variables around the world. An early example in geography was the work of Birkin and M. Clarke (1989), who simulated small-area incomes in Leeds, UK. Such an approach, taking into account the likely income profile of a small area, has the advantage of assessing the kind of people living in an area as well as, crucially, the kind of area it is. Such methods are dependent on the generation of rich representations of the population as individual and household members. This is why spatial microsimulation has often been preferred to array-based demographic representations. Ballas *et al.* (2013) provide income estimates for Kyoto in Japan. This simulation showed pockets of high income in north-east Kyoto and suggested an increasing trend in Japanese cities for social segregation.

In terms of the SIM microsimulation allows a much richer set of variables to be included. Thus, following Birkin *et al.* (2010), rather than the classical approach described by

$$O_i^m = \sum_k e^k P_i^{km} \tag{6.1}$$

in which the population (P), expenditure rates (e) and resulting demand estimates (O) are arrays of limited dimension, the demand model expands to the form

$$O_i^m = \sum_{n \in i} \prod_{m1=1}^{N(m)} x(m1, n) e(m1) \tag{6.2}$$

where $x(m1,n)$ represents a list of binary (zero-one) attributes for a set of n households across a large number of characteristics ($N(m)$), representing diverse social, demographic and household characteristics; $e(m1)$ are specific demand profiles that can be generated from government surveys, market research or retail customer data. For example, Birkin et al. (2002) discuss examples for the estimation of financial service expenditures in relation to diverse factors, including age, gender, household size, dependent children, occupation, income and ethnicity.

6.6 Demand from tourists

In 2013, the UK tourism sector contributed £127bn to UK gross domestic profit (GDP), supported over 1.75 million jobs, and represented one of the fastest growing economic sectors in the UK (Deloitte 2013). Catering facilities are some of the most frequently used by tourists (Ashworth and Tunbridge 1990). Most forms of serviced accommodation, such as hotels, guest houses and bed and breakfast providers (B&Bs) offer some form of catering to guests; whereas, by definition, self-catering accommodation, including camping and caravanning, gives tourists the opportunity to eat out or to purchase and prepare their own food. 'In many holiday locations, during the summer season tourists will complement revenue derived from residents for a range of retailers such as supermarkets, chemists, newsagents, pubs and cafes' (Dudding and Ryan 2000: 302).

The services relied upon by visitors often experience significant fluctuations in visitor utilisation, with many UK tourist resorts identifying peak visitor numbers during the summer months. Peak season visitors are consequently an important driver of demand and expenditure, directly supporting a range of attractions and services that rely on income from tourists. Figure 6.6 shows the extent of seasonal demand in Cornwall, UK in 2012. Commonly, discussion of seasonality is concerned with the under-utilisation of facilities in the low season, which is inefficient, or the excess pressure on facilities during the high-season, which may put pressure on the environment or lead to problems such as congestion. Indeed, retailers note that some of their stores in tourist resorts experience noticeable sales uplift during the tourist season, attracting significant additional revenue originating from outside their residential catchment.

Newing *et al.* (2013b, 2014) provide details on one methodology for estimating tourist spend based on a combination of tourist surveys, backed up in this case by customer data provided by a major UK supermarket.

6.7 Work-based demand

So far the estimations of demand have been based on residential locations of the population. These will still be very important starting points for many retail-based trips, especially at weekends and evenings. However, to estimate mid-week trade (especially 9 a.m. – 5 p.m.) estimating flows of expenditure from places of work and flows from multipurpose trips may also be desirable. The traditional journey made from home to work and back again has, over recent years, mushroomed into a series of multifaceted subtrips (e.g. journey to shop, journey to leisure, journey to school) that play an important part in determining customer behaviour in a number of market sectors (see the work of O'Kelly 1981; Mulligan 1983; and Arentze *et al.* 2005 for more detailed theoretical insights into adding multipurpose trips to location models). Considering work-based locations is especially important in certain geographical locations. Using straightforward census data, retail demand in Canary Wharf, London, for example, would be small as residential demand is low. However, we know that with a large, predominately young professional daytime work population such demand is likely to be considerable. Another good example here would be UK book retailer WHSmith's store in Holborn, Central London, which historically has been one of the top-performing outlets but regarded as lacking potential by a standard model that does not allow for workplace demand. So to reproduce market performance effectively, we need to also consider expenditure originating from the workplace as well as in relation to residential locations. In

Figure 6.6 Seasonal variations in sales for selected grocery stores in Cornwall (£s per week) in (clockwise a–f): (a) winter (Dec.–Feb.) (b) spring (March–May) (c) summer (June–Aug) (d) autumn (Sept.–Nov.) (e) August (peak school summer holidays) (f) 52 week average.

Source: Newing *et al.* (2014)

a SIM framework we can replace the simple O_i term with something more sophisticated (cf. Birkin *et al.* 2010):

$$O_i^1 = \mu_i^1 \sum_k e^k P_i^k \tag{6.3}$$

$$O_i^2 = \sum_j (1-\mu_j^2) \sum_k e^k P_i^k T_{ji} / T_{j*} \tag{6.4}$$

These equations state that demand is split into a residence-based component (n=1), and a work-based component (n=2), in accordance with the parameter μ – which might be calibrated in relation to considerations such as economic activity rates and retail market accessibilities, and then reallocated in relation to commuter flows between origins and workplaces (T_{ij}). We also find that work-based demand is much more strongly spatially conditioned than residential demand ($\beta^2 >> \beta^1$, so we have high β values because people cannot travel far during their lunch hours).

Another important future requirement will be to estimate demand at other types of 'nodes', especially relating to transport networks. As we saw in Chapter 2, a growing interest exists for this type of estimated demand from retailers contemplating airport and railway stations as retail destinations. The journey to work is crucial, again, for the latter. Approximately 62 million passengers pass through Waterloo Station in London each year making it a major retail opportunity.

6.8 Inelastic demand

An issue often ignored in much of the literature within retail location research is dealing with elastic demand (though see Ottensmann 1997). The leisure industry is a good example of a market seemingly able to 'create' new demand once new stores are opened. Modelling the likely impact of creating new out-of-town entertainment facilities for example (bingo, cinemas, ten-pin bowling etc.), means that demand estimates should include latent or potential demand that might be unleashed by new or additional facilities. Given that this demand is effectively money taken from individuals' disposable incomes, demand is not simply created but switched, most likely from other leisure pursuits in a local area (e.g. pubs and fast food outlets). The quantification of the level of switching by small-area and customer type provides the key to handling this elastic element in a constrained way. How can this type of demand be modelled? Again here we follow the discussion in Birkin *et al.* (2010):

In the basic model, we have

$$O_i = \sum_k e^k P_i^k \tag{6.5}$$

One way to measure elastic demand is to first compute an accessibility index H_i (such as that provided by Hansen 1959) as a measure of the variations of availability in space:

$$H_i = \sum_j W_j exp^{-\beta c_{ij}} \tag{6.6}$$

This quantity allows 'average' demand to be adjusted as follows

$$O_i = f(H_i) \sum_k e^k P_i^k \tag{6.7}$$

where $f(H_i)$ would typically take the form of a logistic function, in which the parameters of minimum adjustment, maximum adjustment and the acceleration of demand with respect to accessibility can be calibrated to individual market conditions.

Here, as in the cases that follow later in the book, the adoption of equations (6.5–6.7) supersedes the classic entropy-maximising assumption that demand is fixed.

6.9 Future demand

In the rest of this chapter we focus on issues that are likely to significantly impact demand estimations in the future. Current demand is of course vital to understand market performance or potential but what will that demand look like in 10 or 20 years' time? Long-term stability (or better still growth) is important to guarantee future success. The retail market will change in important ways over the next 10 to 20 years. We discuss these changes under three headings: the geography of population change; the growing demand for convenience; and the importance of 'new' consumer groups to offer retailers new opportunities for growth.

6.9.1 The geography of population change

Within all countries, cities and regions, population change is important to understand and predict. Figure 6.7 shows the growth and decline of UK regions between 2001 and 2011.

Figure 6.7 shows areas of substantial growth in the UK – areas that are providing considerable new opportunities for retail development. This helps to support arguments that retail saturation can only be considered in a spatial context, taking into account demand-side changes (Guy 1996; Langston *et al.* 1997; Lord 2000: interestingly, Maraschin and Krafta 2013 have constructed a retail growth model based on levels of saturation and population growth). The major growth area can be seen in the 'home counties' (a

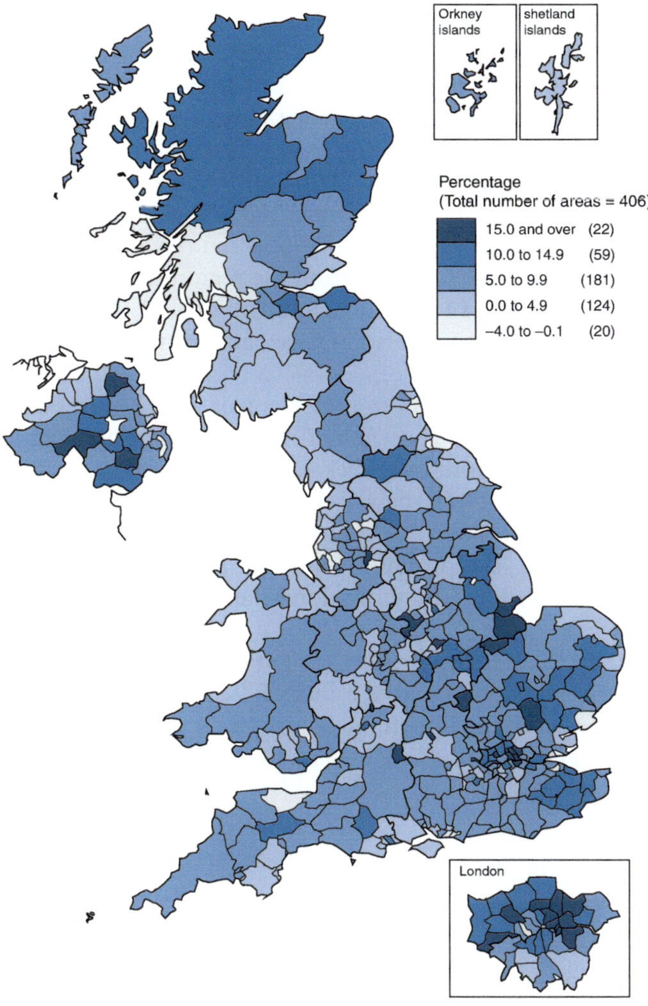

Figure 6.7 UK population change 2001–11.
Source: ONS (2012a)

swathe of commuting areas around London, especially to the north), which has offered retailers considerable potential for growth since 2001. Some of these regions will continue to grow significantly over the next 20 to 30 years. Also, many inner-city areas have seen sustained investment in new housing, particularly aimed at young professionals (under a regeneration or gentrification banner): east London, Manchester and Leeds for example). In contrast, the maps show decline in many urban areas of the north of England (former manufacturing towns such as Barrow in Furness, Sunderland,

Middlesbrough, Burnley and Stockport). ONS (2012b) predict much of the highest growth between 2011 and 2020 will take place in London and the south-east, especially around Luton, Bedford and Cambridge for the latter. It is crucial for retailers to understand these population dynamics in relation to future investment opportunities. While the growth areas offer opportunities for the retailers, the areas of decline show areas to avoid, or perhaps to be wary of, when considering longer-term dynamics. To help identify these areas Debenham *et al.* (2003a, 2003b) demonstrate that it is possible to identify 'vulnerable' areas based on the percentage of jobs that rely on a few key employers. If these companies were to divest from the region then considerable retail difficulties may follow. In addition, retailers should understand not only population change but changes also in social and economic disparities over time (see Lloyd 2016 for example).

Part of any future population growth is the extent of new house build. Although in the UK the last Labour government's plans for new housing in four major growth areas have been officially shelved (they were devised before Labour lost the UK election in 2007), these four areas are still likely to see significant house build programmes in the future. These include the Milton Keynes/south midlands area, the London–Stansted Airport–Cambridge corridor, the Thames Gateway (east London along the Thames River) and Ashford in Kent.

Growth figures in the UK are also driven by high rates of immigration. During the 2000s many regions saw a large influx of East European immigrants. Figure 6.8 shows this change in ethnic composition in the UK since 2001. It is likely that it is already providing new retail opportunities and will cause further growth in these regions in the future (see discussions of ethnic demand in Section 6.3).

6.9.2 Changing household structures and the demand for greater convenience

In most developed countries of the world family size and hence household size have been declining over time. At the same time we have witnessed a rise in the number of single-family households, as individuals put off marriage to a later date (especially young professionals) and more elderly persons are surviving longer but often without their spouse. Gofton (1995) provided evidence of the impact of this type of changing household composition on grocery and food retailing behaviour. He also suggested an increase in dual-income households has resulted in an increase in consumption of convenience foods. De Kervenoael *et al.* (2006) also discuss the importance of this behavioural shift in which time has seemingly become a more precious commodity. However, they also highlight the importance of the increase in single-person households and the corresponding shift to a convenience culture. This rise in the demand for convenience is often most strongly associated

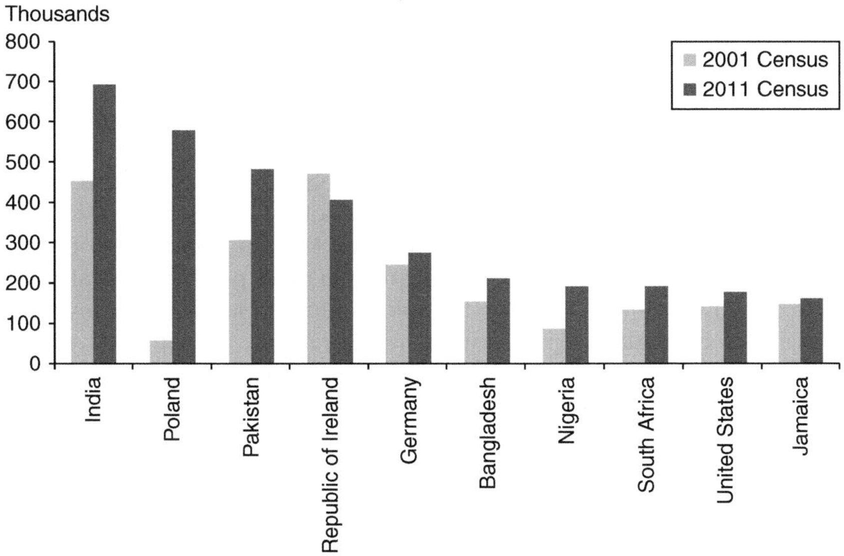

Figure 6.8 UK ethnic change 2001–11.
Source: ONS (2012c)

with young, single-household professionals who work long hours, eat out regularly and are less inclined to cook for themselves (the 'money-rich, time-poor' consumers).

There have been a number of large-scale UK studies of retail behaviour over time that also suggest convenience retailing is becoming more popular. In a study of Portsmouth residents, I. Clarke *et al.* (2006) and Jackson *et al.* (2006) report that from 1980 to 2002 food shopping behaviour had become increasingly dependent on the geographic proximity of retail units and that convenience had come to play a key and increasing role in shopping. Research conducted by Hallsworth *et al.* (2010) also found evidence of an increase in shopping frequencies among consumers in the UK retail market; from 1980 to 2002, the proportion of participants in their survey food shopping three times a week increased from 9 per cent to 21 per cent, and convenience stores are often ideally placed to accommodate this behaviour.

Thus, it can be argued that a combination of demographic and lifestyle shifts have created an increased demand and utilisation of convenience retailing by consumers in UK. A number of retailers have been happy to feed, and indeed, fuel this demand through diversifying store portfolios to be move heavily reliant on convenience retailing (see again the supply-side changes discussed in Chapter 2).

6.9.3 'New' consumer groups

Population ageing is one of the major issues facing developed worlds. The implication for planning of all types of services is massive. Table 6.3 shows the predicted increases in the UK elderly population up to 2037.

The increase in the elderly population will have a number of impacts as far as retail demand is concerned (the chase for the 'grey pound'). First, the nature of products sold may change considerably. The most obvious examples may be wheelchairs, stair lifts, motor scooters etc. (see, for example, the rise in the USA of the stores of 'World of Mobility' and in the UK of 'Allmobility'). More subtly, we will see changes in other products. Royal Town Planning Institute (RTPI 2004) for example, note the potential for many more local health food outlets, representing in part a response to demand by older people for more 'healthy eating', dietary supplements ('nutraceuticals') and herbal remedies. In addition, many of these new pensioners will be relatively well-off, perhaps having inherited property from their parents. These retail clients may thus have low or no mortgage payments and high disposable incomes. Polk in the USA, for example, include such a group in their lifestyles database: 'nomadic grandparents' are those affluent pensioners with a penchant for cruises and extended holidays. Some companies are also beginning to change their marketing campaigns around new products for the elderly – Coca-Cola, for example, targeting its former customers with new wines and coffees which they perceive as being more attractive drinks to the over 65s.

In terms of general impacts RTPI (2004: 42) interestingly suggest:

> "The ageing population profile may not affect the distribution of retail activity significantly, except in the longer term when an increasingly elderly population will require more local, walkable shopping provision. Demand for new retail locations should slow as an older population spends relatively less on products, as opposed to services, and some growth in local convenience food shopping seems likely. Neighbourhood '7 to 11' stores, rather than 'retail shed' developments will help serve this demand."

Table 6.3 Projected population by age in millions, UK, mid-2012 to mid-2037.

Ages	*2012*	*2017*	*2022*	*2027*	*2032*	*2037*
0–14	11.2	11.7	12.2	12.3	12.2	12.2
15–29	12.6	12.4	12.1	12.3	12.9	13.3
30–44	12.8	12.7	13.3	13.6	13.5	13.2
45–59	12.6	13.3	13.0	12.6	12.4	13.0
60–74	9.4	10.1	10.7	11.6	12.3	12.1
75 and over	5.0	5.5	6.6	7.7	8.5	9.5
75–84	*3.6*	*3.8*	*4.6*	*5.3*	*5.4*	*5.9*
85 and over	*1.4*	*1.7*	*2.0*	*2.4*	*3.1*	*3.6*

Source: ONS (2013)

What will be the spatial implications of population ageing? First, the growth in pensioners will not be felt evenly across any country. For example, ONS (2012d) estimate that UK coastal areas and retirement country hot spots will obviously gain the most. Figure 6.9 shows the current location of over 65s in England and Wales and the coastal pattern is clear to see. Retailers should be gearing up to considerable increases in the elderly in these areas.

Another consumer group likely to increase in numbers and hence produce a new type of retail demand is the gay community (the 'pink pound': the group as a whole is often referred to as GLBT: gay, lesbian, bisexual and transsexual)). Estimates of the size of the gay population vary widely. In the

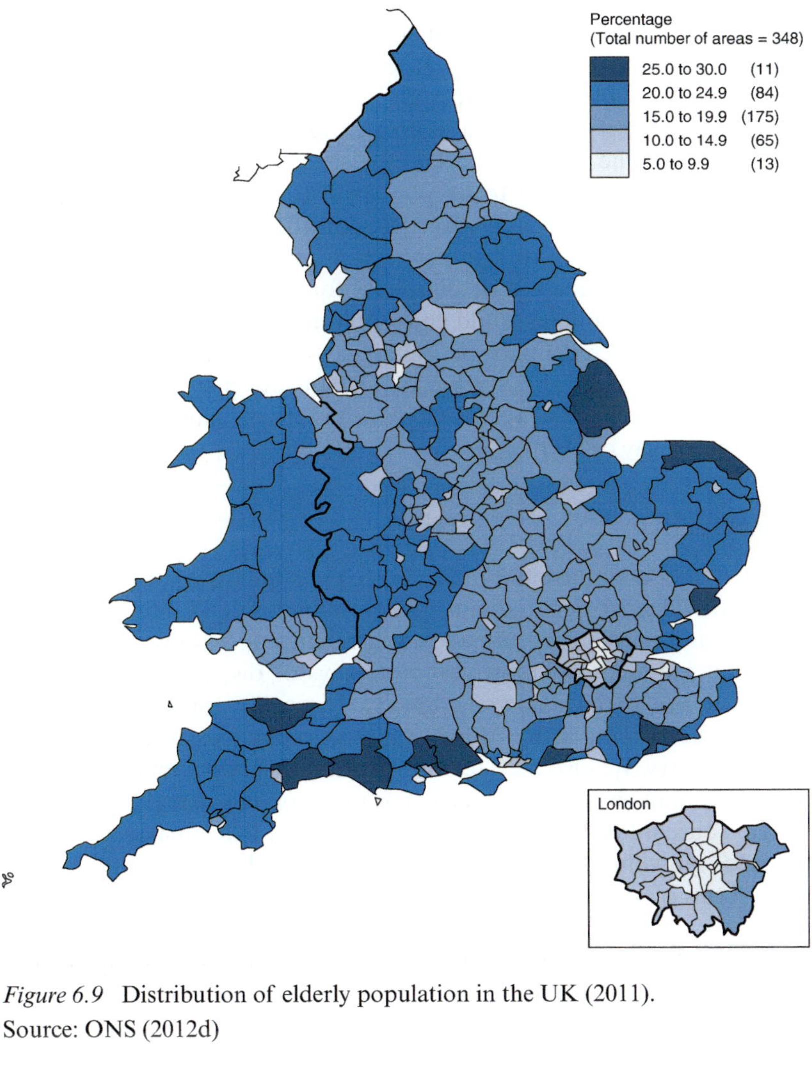

Figure 6.9 Distribution of elderly population in the UK (2011).
Source: ONS (2012d)

UK it is estimated to be around 3.5 million with a combined annual spending power of £100 billion, while in the USA estimates suggest 180 million and an estimated spending power in 2014 of around $880 billion (Witeck Communications 2015). Given such spending power, it is not surprising that there is a growing literature on marketing to the gay community (see for example Peñaloza 1996; Oakenfull and Greenlee 2005) and studies into gay consumer behaviour (Kates 2000, 2013; Schofield and Schmidt 2005, for example). There is also a clear spatial pattern to this demand. In the UK notable gay communities exist in Brighton, London, Manchester, Blackpool and Bournemouth. For more discussion of the geography of the gay community in the UK, for example, see Aspinall (2009). In the USA, according to the latest census results, the top ten cities for same-sex couples are: 1. Fort Lauderdale, Florida; 2. Seattle, Washington 3. San Francisco, California; 4. Minneapolis, Minnesota; 6. Portland, Maine; 6. Somerville, Massachusetts; 7. Oakland, California; 8. Providence, Rhoda Island; 9. Washington, DC; 10. Warwick, Rhonda Island. Whittemore and Smart (2016) offer an interesting methodology for identifying small-area concentrations of the gay community in the USA through the study of local advertisements.

The gay community is said to be very loyal to retail firms that encourage gay consumers either through advertising or offers on products (Business Insider 2013). For example, Budget cars saw a large increase in sales in the USA following its introduction of discounts for gay couples (matching the discount that married couples could enjoy on renting cars). Other US retailers with well-known credentials of being gay-friendly include JCPenney, American Airlines, Target and Starbucks. The reasons may be varied – supportive of gay rights, gay-friendly workplace practices, marketing using gay advertisements or gay celebrities etc.

6.10 Conclusions

This chapter first examined the many ways of estimating demand. It has shown that the idea of demand can be deceptively simple. For most retail products demand needs to be considered in relation to who is the principal target for that product. The most common ways of disaggregating demand include gender, age, social class and ethnicity. These different socio-economic variables can be combined by using geodemographics or microsimulation models as necessary. We also need to consider how that demand might change in the future both in terms of how the market is changing (population ageing, more single households, more gays/lesbians) and how this impacts on the size and nature of individual store catchment areas. Some locations will continue to grow in terms of size; some will grow but change very rapidly as the population changes; some will decline as population leaves. An understanding of the dynamics of the marketplace is a key ingredient in undertaking store location research.

7 Measuring the attractiveness of retail stores or shopping centres

7.1 Introduction

In Chapter 6 we discussed the methods for estimating consumer demand and the ways in which demand is likely to change in the future. In this chapter, we introduce the supply-side of the retail interaction process, examining different methodologies for estimating store or shopping centre attractiveness. For any site location methodology deciding what makes stores or shopping centres attractive is important. In terms of the spatial interaction models (SIM) introduced in Chapter 5, store or centre attractiveness is one of three key variables, captured through the W_j term. This chapter will focus on a wide variety of issues relating to store attractiveness and consumer choice of destination. The rest of the chapter is set out as follows. In Section 7.2 we discuss the nature of retail destinations, exploring what differentiates a single store from a larger shopping centre. In Section 7.3, we review the different variables used to measure store attractiveness as seen in the literature, and in Section 7.4 we introduce further issues that need to be considered in relation to store choice. Here, the discussion in Chapter 6 becomes relevant again as we bring demand and supply issues more formally together. These issues include brand choice in relation to consumer type, brand loyalty, the impacts of the so-called network effect (the impact that a set of stores can have in influencing regional market share) and finally, the concept of hierarchical destination choice.

7.2 The nature of retail destinations

Several important choices regarding the representation of supply points have to be made first in retail location modelling. Whether facilities are identified as stand-alone outlets, or grouped into centres, is the first crucial decision. For many retail activities – automotive, petrol and supermarkets are all obvious examples – representing flows to individual outlets seems most appropriate. More problematically though is the fact that the distinction between outlet-based systems and centre-based systems is becoming increasingly blurred. The incorporation of 'stand-alone' retail facilities into retail parks often means that such locations are more akin to centres than stand-alone outlets (see Thomas *et al.* 2004) – the presence of both competitors and complementary retailers adding to the overall 'attractiveness' of an outlet. Similar developments are observable in other sectors. The emerging trend towards agglomerations of activity in automotive distribution (the 'automall' in the USA) and the proliferation of multifranchised outlets is transforming choosing a new car into something more akin to comparison shopping.

For most retailing activities, we find the definition of 'centres' to be helpful and necessary. The clustering of outlets allows for the existence of genuine scale economies within a retail centre (business typically is better in a large centre than in a smaller neighbourhood centre). These larger centres typically have better facilities, from car parking to restaurants, and enjoy safer, warmer, drier environments – all increasing their attraction to customers. The impact of retail adjacencies can also be crucial at both the local and centre level. For example, many note the importance of being close to a major anchor store such as Marks and Spencer in the UK or JCPenney in the USA. Whether scale economies are always beneficial to individual outlets within a retail centre is a moot point, and may, in part, be determined by levels of saturation in local markets, as we explain in detail in the sections below.

Also of note is that centre definition is not in itself an unambiguous process. Where does one centre end and another begin? Which sub-centres need to be explicitly identified? For example, there is no determinate answer to the question of 'how many retail centres are there in the UK'? In work undertaken on behalf of the UK Post Office, we analysed more than 11,000 branches, often represented both within and outside conventional retail centres (see Chapter 9 for more details). Here 'fuzzy' retail centre membership was adopted so that outlets close to a major retail centre might experience some of the benefits of retail adjacencies and centre quality, but not necessarily on the same scale as other outlets more clearly located within the core of the retail centre (see also the discussion in Section 7.4.4).

7.3 Measuring store/centre attractiveness

This section provides a review and reassessment of the measurement of the attractiveness of individual stores or centres. The production-constrained SIM, as described in Chapter 5, is typically used to estimate shopping revenues based partly on the supply-side variable W_j. This variable represents the attractiveness of destinations. Many factors can be seen to affect the attractiveness of retail stores and it is therefore necessary to devise the best method of calculating a destination's attractiveness. A consumer's choice of destination will be dependent on both individual store characteristics and the characteristics of the centres in which they are located, and therefore various elements of potential attractiveness have to be taken into account.

Vickerman (1974) states that size and turnover, as used to calculate attractiveness in the early retail models of Huff (1963, 1964) and Lakshaman and Hansen (1965), do not adequately measure the real elements of attraction because other factors also influence destination choice. However, store size or centre size is most frequently used as the attractiveness term in SIMs because they are easy to measure and are often assumed to be the best proxy for other measures of attractiveness (Pacione 1974).

However, intuitively, we know that store choice by consumers is dependent on more than store size and ideally all store and centre characteristics that may have an influence on shopping behaviour should be included in any attractiveness variable. A range of both objective and subjective measures have been used in previous studies to calculate the attractiveness of destinations. 'Objective' factors (objective in the sense that they are quantifiable) such as store size, parking facilities and relative prices have been regularly included in site location models. 'Subjective' measures such as the consumers' perception of brand image and customer service have also been used to measure attractiveness. Fotheringham and Trew (1993) used a household survey and a logit model to conclude that chain image is at least as important as store size and competition in determining destination choice. McCarthy (1980) undertook a multinomial logit analysis which indicated that generalised shopping centre attributes, as derived from attitudinal information, are significant in determining destination choice behaviour. Oppewal *et al.* (1997) undertook a study using conjoint analysis and multinomial logit models in order to determine centre characteristic factors important in influencing destination choice for food retailing and for clothes and shoes. They discovered that although the size of the retail destination was the most important factor determining centre choice, other variables such as centre location (convenience) and physical appearance and layout of centres were also influential. The available selection of competing stores and the mix of types of store in the centre were also significant determinants of a centre's attractiveness to consumers. Pacione (1974) found, using stepwise regression techniques, that the presence of other store types, such as supermarkets, department stores and banks, was the second most important variable (after distance) in the determination of consumer travel patterns. Another centre characteristic found to be important in the determination of centre attractiveness is parking facilities. In a Dutch study using a multiattribute model (containing many attractiveness variables: parking, number of shops, variety, number of employees, number of stores and distance) parking, after distance, was the most significant variable (Timmermans 1981). More recently Henriques-Marques *et al.* (2016) explore the importance of 'atmospherics', which include factors such as store design and the pleasantness of the atmosphere and might include empathy with the staff.

Therefore, simply using store size to measure destination attractiveness may not reflect all the factors that influence destination choice in the real world. Spencer (1978) has argued that it is possible to improve the performance of models of shopping behaviour by using variables that are more closely related to consumers' perceptions of shopping places than the traditional attractiveness variables such as store size. However, the inclusion of all the variables that affect destination choice is often difficult due to the inadequacy of the data. It is especially difficult to estimate the subjective measures of attractiveness as this involves the use of consumer survey information. Therefore, it is necessary to decide which outlet characteristics must be included in the attractiveness measure and perhaps only deal with those.

To illustrate the fact that a more complex and in-depth study of attractiveness can improve the explanatory power of a model we now summarise the work of Eyre (1999). She worked with a major UK high street retailer to improve the fit of a SIM where the attractiveness term was first built around store size and brand attractiveness only (called the 'old' model here). The study area was the Yorkshire TV region. There were 29 shopping centres in the Yorkshire TV region at that time that contained one or more of the client's stores; these are shown in Figure 7.1. The goodness-of-fit statistics for the applied model (the old model) for these 29 centres are shown in Table 7.1.

Table 7.1 shows that the aggregate (old) model did provide a good model fit overall. The model fit is acceptable for an applied retail location model but there were some notable outliers – centres that were modelled poorly in an otherwise well-fitting model. The aim of the study was to try to discover factors that would help to understand why some centres were over or under-predicted in terms of revenue in the model. To summarise the extent of the outliers a 'centre performance' score was derived by comparing the observed and predicted revenues for all client stores, for all goods within each centre. The centre performance factor was calculated as:

$$\text{centre performance} = (\text{observed centre revenue}/\text{predicted revenue})*100 \tag{7.1}$$

The centre performance levels for the 29 Yorkshire centres, ranked with the highest performing centres at the top, are shown in Table 7.2. The scatter plot shown in Figure 7.2 also represents the differences in performance between centres for the model, with centres below the (imaginary) 45 degree line over-performing (being under-predicted in the model) and those above the line under-performing (being over-predicted in the model).

It can be seen from Table 7.2 and Figure 7.2 that there was significant variation in performance levels between centres. Table 7.2 also indicates that the degree of fit for centres was not simply dependent on their size. Although Meadowhall (a regional shopping centre in Sheffield) and York were over-performing compared to model estimations, small centres such as Ilkley and

Table 7.1 Goodness-of-fit statistics for the old model.

Good	*SSE*	r^2	r_s
All goods	43628664	0.89	0.97

Source: Eyre (1999)

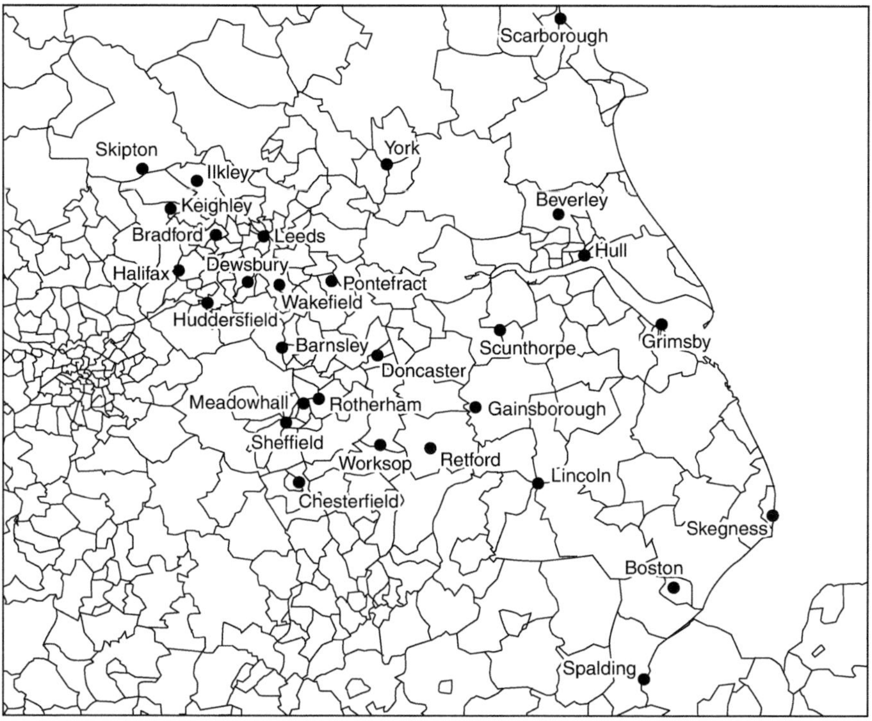

Figure 7.1 Centres in the Yorkshire TV region containing client stores.
Source: Eyre (1999)

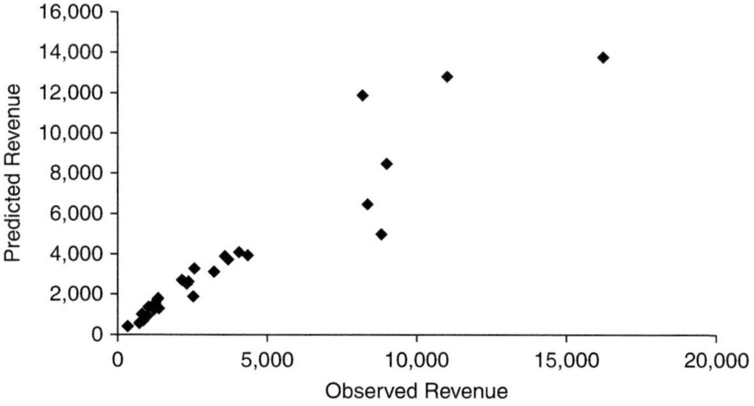

Figure 7.2 Scatter plot of observed and predicted centre revenues for the old model.
Source: Eyre (1999)

Table 7.2 Centre performance levels for the old model.

Centre	*Observed revenue*	*Predicted revenue*	*Centre performance*
Meadowhall	8,822	4,980	177.2
Scarborough	2,530	1,886	134.1
Ilkley	733	557	131.5
York	8,360	6,460	129.4
Skipton	863	692	124.6
Leeds	16,215	13,764	117.8
Lincoln	4,351	3,939	110.5
Boston	1,387	1,284	108.0
Beverley	1,003	935	107.3
Hull	8,997	8,469	107.2
Wakefield	3,223	3,123	103.2
Huddersfield	4,054	4,087	99.2
Dewsbury	1,229	1,244	98.8
Grimsby	3,685	3,737	98.6
Spalding	1,198	1,293	92.6
Halifax	2,318	2,509	92.4
Doncaster	3,581	3,894	92.0
Scunthorpe	2,377	2,645	89.9
Rotherham	1,188	1,350	88.0
Sheffield	11,028	12,800	87.2
Gainsborough	338	408	82.8
Retford	818	993	82.4
Skegness	1,263	1,543	81.8
Keighley	2,160	2,707	79.8
Barnsley	2,566	3,279	78.2
Pontefract	1,341	1,744	77.9
Chesterfield	1,040	1,360	77.4
Worksop	1,364	1,786	77.3
Bradford	8,191	11,871	69.0

Source: Eyre (1999)

Skipton also seemed to be over-performing, while some large centres such as Bradford and Sheffield (city centre) were seen to be under-performing by a large amount.

There could be several reasons for such variations in model performance. In the old model no consideration was given to centre attractiveness: only store size and brand were in effect used. The client's store in Meadowhall, for example, received an actual revenue nearly double that expected by the model, given its size and the attraction of other retailers in the large shopping centre. This centre is a large, out-of-town, regional shopping centre that attracts people from long distances. It is also close to the M1 motorway making access much easier than Sheffield city centre, especially from the north and south. The success of this store could be due to shoppers going to this centre to take advantage of the large amount of shopping floor space that is available for comparison goods such as clothing and shoes. The client's store may not be the primary destination of the customer in Meadowhall, but rather the store benefits from the attractiveness of the centre in terms of other store types. There are other factors, therefore, at work besides the size and brand of the client store in the centre.

Other centres were also over-performing but this could be due to different reasons. For instance it can be seen that the centre of Scarborough was under-predicted by the (old) model by 35 per cent. This could be due to the large amount of tourism that occurs in the town. Tourism could also be a partial explanation in the under-prediction of centre revenues for York, Ilkley and Skipton. The centre that was most significantly over-predicted by the model was Bradford. Bradford contains lots of independent shops, many of which are low-quality bargain or charity shops, and much of the city centre is not pedestrianised or protected from the poor weather. This could explain why the city does not attract as much revenue as it perhaps should, given its size. The under-performance of centres such as Worksop, Pontefract and Barnsley could also be explained through this factor (don't forget that the demand is estimated by the O_i term in the SIM so variations in expenditure available between areas had already been accounted for).

The task was thus to try to formulate a centre attractiveness function that would remove or lessen these differences in performance. Such a function could be made up of several factors. Eyre (1999) chose the following variables to measure and include in the (new) model:

1 Provision of parking facilities.
2 The number of department stores.
3 The number of banks and building societies.
4 The proportion of stores in each centre that were retail multiples (firms who own more than ten outlets nationally). This factor is based on the hypothesis that multiples will be more attractive to consumers and therefore centres with a large percentage of retail multiples will be more attractive.
5 The percentage of stores that are in undercover centres or in pedestrianised areas.

Information concerning the types of shops in centres and parking facilities was extracted from UK Goad city centre plans and site visits. These maps provide a detailed breakdown of which shops are located in which centres. The plans also provide information concerning car parks, including the number of spaces in each car park. These maps can be used to find out how many key department stores, banks and parking spaces etc. are present in centres, and also provide the information required to calculate the proportion of retail multiples in each centre. The data collected for the 29 chosen centres are shown in Table 7.3. The centres are again listed in order of performance, with the best performing centres at the top. The five variables were scored using a scale of 1–5 and these were subsequently added together to give a 'field survey attractiveness' factor with a maximum score of 25.

Table 7.3 The observed attractiveness factors for centres.

Centre	*Parking spaces*	*% multis*	*% covered/ pedestrian*	*Banks/ building societies*	*Dept stores*	*Field survey attractiveness*
Meadowhall	12,000	77	100	6	7	25
Scarborough	1,806	44	60	18	6	22
Ilkley	764	35	10	11	5	21
York	643	50	30	27	12	24
Skipton	840	33	15	10	5	21
Leeds	4,680	57	60	40	15	22
Lincoln	2,224	51	48	31	8	21
Boston	2,090	32	21	15	7	17
Beverley	590	54	56	15	2	19
Hull	3,019	56	59	41	13	19
Wakefield	2,343	50	57	20	5	17
Huddersfield	2,669	41	28	25	7	14
Dewsbury	1,890	32	18	10	4	9
Grimsby	1,953	47	60	19	6	14
Spalding	668	30	22	13	6	12
Halifax	1,504	33	42	19	5	13
Doncaster	2,530	34	25	27	11	16
Scunthorpe	2,059	28	34	14	8	11
Rotherham	1,578	39	30	16	3	10
Sheffield	4,097	40	37	44	11	11
Gainsborough	385	25	31	8	4	11
Retford	642	30	45	10	2	15
Skegness	227	30	16	8	5	13
Keighley	1,670	44	46	15	6	10
Barnsley	2,320	48	34	23	6	10
Pontefract	1,150	38	33	11	5	8
Chesterfield	1,741	36	27	20	7	11
Worksop	1,066	26	34	12	4	8
Bradford	1,820	28	21	27	9	10

Source: Eyre (1999)

A logit analysis was carried out to test which of the above attractiveness factors were important in determining centre performance. As noted in Chapter 5, the logit model is a form of generalised linear model that allows the formation of a relationship (similar to multiple regression) between a categorical dependent variable and several independent variables. In this analysis, store (centre) performance is the dependent variable and the centre attractiveness factors are the independent variables. Correlations were also undertaken in order to analyse the relationship between the attractiveness factors and the performance variable. The results of the logit analysis are shown in Table 7.4 and the correlation analysis in Table 7.5.

Tables 7.4 and 7.5 show that the field survey attractiveness score is the most important determinant of variations in centre performance, causing the largest

Table 7.4 Results of the logit analysis to test the importance of different attractiveness factors in determining centre performance.

Attractiveness factor	*Deviance*	*Degrees of freedom*	*Null deviance – model deviance*
Null model	38.50	28	
Parking spaces	31.65	25	7.85
% multiples	21.27	25	17.23
% covered	27.81	26	10.69
Banks/building societies	37.28	25	1.21
Department stores	37.28	26	1.21
Field survey attractiveness	20.72	25	17.77

Source: Eyre (1999)

Table 7.5 Results of the correlations between attractiveness factors and performance for centres in the Yorkshire TV region containing client stores.

Attractiveness factor	*r*	r^2
Parking spaces	0.67	0.45
% multiples	0.67	0.45
% covered	0.54	0.29
Banks/building societies	−0.11	0.01
Department stores	0.16	0.03
Field survey attractiveness	0.83	0.69

Source: Eyre (1999)

decrease in deviance for the logit model and having the highest value of r^2. This is followed by the percentage multiples factor, the number of parking spaces and the percentage of stores that are undercover or in a pedestrianised area. The other variables examined were found not to be significant at the 95 per cent level.

Therefore the centre attractiveness score was used to replace the current attractiveness value in the (new) model. The goodness-of-fit statistics for this new model are shown in Table 7.6 and the new revenue predictions and performance indicators for the 29 centres are shown in Table 7.7.

Through a comparison of Table 7.6 and Table 7.1, it can be see that the incorporation of the new attractiveness term has significantly improved the performance of the model. For all goods, the sum of squared errors (SSE) has decreased by 85 per cent from the old model and the r^2 value has also increased significantly from 0.89 to 0.98.

Figure 7.3 also indicates the improvement of the centre revenue predictions caused by the inclusion of the new attractiveness term. By comparing Figure 7.3 with Figure 7.2, it can be seen that there is less variation away from the 45 degree line in Figure 7.3; the scatter plot showing revenue predictions for the new model. Table 7.7 also indicates that model performance improved

Table 7.6 Goodness-of-fit statistics for the new model.

Good Type	*SSE*	r^2	r_s
All Goods	6,715,813	0.98	0.97

Source: Eyre (1999)

Table 7.7 New centre revenue predictions and centre performance for the new model.

Centre	*Observed revenue **	*Predicted revenue **	*Centre performance*	*% Improvement in performance*
Meadowhall	8,822	8,940	98.7	75.9
Scarborough	2,530	1,915	132.1	2.0
Ilkley	733	515	142.4	−10.9
York	8,360	6,793	123.1	7.3
Skipton	863	668	129.2	−4.6
Leeds	16,215	15,447	105.0	12.8
Lincoln	4,351	4,078	107.7	3.8
Boston	1,387	1,252	110.8	−2.8
Beverley	1,003	917	109.3	−2.0
Hull	8,997	8,853	101.6	4.6
Wakefield	3,223	3,218	100.2	3.0
Huddersfield	4,054	3,939	102.9	−2.1
Dewsbury	1,229	851	144.4	−43.2
Grimsby	3,685	3,704	99.5	0.9
Spalding	1,198	1,063	112.6	−5.2
Halifax	2,318	2,444	94.8	2.4
Doncaster	3,581	3,611	99.2	7.2
Scunthorpe	2,377	2,452	97.0	7.1
Rotherham	1,188	914	130.0	−18.0
Sheffield	11,028	10,982	100.4	13.4
Gainsborough	338	328	103.1	14.1
Retford	818	933	87.7	5.3
Skegness	1,263	1,374	91.9	10.1
Keighley	2,160	2,604	82.9	3.1
Barnsley	2,566	2,796	91.8	13.6
Pontefract	1,341	1,258	107.6	17.5
Chesterfield	1,040	1,174	88.6	12.2
Worksop	1,364	1,301	104.8	18.9
Bradford	8,191	9,770	83.8	14.8

Source: Eyre (1999)

Note: * Revenue in £1,000s

for 21 of the 29 centres and that there was only one centre whose performance decreased significantly, Dewsbury. Several of the centres that were being poorly predicted such as Meadowhall and Bradford are now being modelled much more accurately. Therefore, this is an indication of the improved

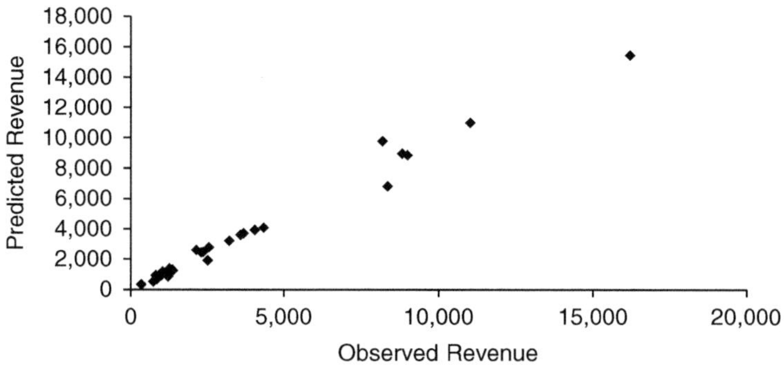

Figure 7.3 Scatter plot of observed and predicted centre revenues for the new model.
Source: Eyre (1999)

explanatory and predictive power of the model subsequent to the addition of the new attractiveness term.

We hope this case study has therefore been useful for understanding the importance of including both *store* and *centre* attractiveness into the supply-side variables in a model.

7.4 Modelling store attractiveness by person type and retail brand

7.4.1 Disaggregating by brand and person type

This section begins to more formally link a number of the demand issues we discussed in Chapter 6 with supply issues from this chapter so far. First, we wish to explore how it is possible to handle some of the more complex and individualised behaviour of different groups of consumers and to take account of key socio-economic characteristics that drive expenditure and store choice. To do this we follow the discussion of Newing *et al.* (2015) and use the results from a detailed consumer survey provided by Acxiom Ltd, and analysed extensively by Thompson *et al.* (2012).

It is recognised that the characteristics of demand and the attractiveness of the retail destination will vary according to a person's income, age, ethnicity or other socio-economic characteristics of the consumer, and may also vary depending on the type of product in question. To add brand choice explicitly into the model framework we need to refine the attractiveness term. This can be done in a number of different ways. For example, as equation 7.2 shows, a power function (α^{kn}) can be incorporated within the attractiveness term in order to apply a measure of relative brand attractiveness to the existing attractiveness term on a consumer-by-consumer basis. The new model can be written as:

$$S_{ij}^{k} = A_i^k \, O_i^k \, W_j^{\alpha^{kn}} \, exp^{(-\beta^k c_{ij})} \tag{7.2}$$

Where: S_{ij}^{k} represents the predicted expenditure flow between zone $_{i}$ and store $_{j}$ by consumer of type k.

A_i^k is the usual balancing factor which takes account of competition and ensures that all demand from zone $_{i}$ by consumer type k is allocated to stores within the modelled region. The balancing factor thus ensures that:

$$\sum_j S_{ij}^{k} = O_i^k \tag{7.3}$$

It is calculated as:

$$A_i^k = \frac{1}{\sum_j W_j^{\alpha^{kn}} exp^{(-\beta^k C_{ij})}} \tag{7.4}$$

O_i^k is a measure of the demand or expenditure available in demand zone i by consumer of type k.

W_j reflects the overall attractiveness of store j, while α^{kn} represents the additional or perceived relative attractiveness of store j for consumer type k and by store type n (often reflecting scale economies).

C_{ij} is the distance between zone i and store j, and incorporates the distance deterrence/decay parameter $exp^{-\beta^k}$ for household of type k.

The model takes the same form as the classic production-constrained SIM introduced in Chapter 5, yet the balancing factor (A_i) demand (O_i) supply (W_j) and distance deterrence ($exp^{-\beta C_{ij}}$) terms have been modified to incorporate different behaviours by consumer types (k). The inclusion of these additional terms allows both supply and demand to be disaggregated independently, yet the links between them maintained through the recurrence of consumer type (k) on both the demand and supply-side of the equation.

This disaggregation by both consumer type and retailer brand affords tremendous potential for the model to incorporate flows between different consumer types and different retailers, through modified attractiveness and distance terms. Thompson *et al.* (2012) use Acxiom's research opinion poll (ROP) (2009 and 2010 data) in combination with the official UK government's 'Output Area Classification' (OAC) geodemographic system (Vickers and Rees 2007; see also Chapter 4) to identify each retailers' customer base. They create a location quotient for each retailer, dividing that retailers' observed customer breakdown by OAC group by the underlying distribution of population across the OAC groups in their study region. As such, these location quotients identify whether a particular OAC group is over or under-represented in a retailers' customer profile. They note, for example, that in the

UK Waitrose, Marks and Spencer, and to some extent Sainsbury's, all generate high patronage from the affluent 'city living' supergroup (most likely to be young professionals), while the same is true of Asda in the 'blue collar communities' supergroup (typically manual workers), Co-op in the 'countryside' supergroup and Sainsbury's in the 'prospering suburbs' supergroup.

The location quotients produced by Thompson *et al.* (2012) have been used here to inform the use of additional brand attractiveness, via the alpha parameter, for each retail fascia and each consumer group. The location quotients have been rescaled around the value of 1, since alpha operates as a power function on store attractiveness in the model. As such, store floor space, for example, is raised to a power, depending on the individual combination of customer type and store brand/fascia, thus recognising that a unit of floor space of Waitrose is more attractive than a unit of floor space of Asda to certain household types. The rescaled location quotients are shown in Table 7.8.

The alpha parameter is intended to control the relative attractiveness of different brands to different household types, based on geodemographic type. Following the introduction of alpha as a model parameter we would expect higher end retailers, such as Marks and Spencer, Waitrose and Sainsbury's to be more attractive to high-income households and less attractive to low-income households, while discount retailers (such as Lidl, Aldi, Iceland and, to an extent, Asda) to be relatively more appealing to low-income households. Therefore

Table 7.8 Brand location quotients for use in disaggregated SIM.

Brand (retailer)	*OAC supergroup*						
	1	*2*	*3*	*4*	*5*	*6*	*7*
	Blue collar	*City living*	*Countryside*	*Prospering suburbs*	*Constrained by circumstances*	*Typical traits*	*Multicultural*
Aldi	0.9980	0.9970	1.0051	0.9987	1.0025	1.0005	0.9952
Asda	1.0076	0.9912	0.9904	0.9970	1.0023	0.9992	1.0013
Co-op	1.0020	0.9990	1.0157	0.9922	1.0008	1.0000	0.9894
Lidl	1.0015	0.9995	1.0066	0.9962	0.9957	0.9997	1.0091
M&S	0.9891	1.0381	0.9967	1.0066	0.9952	1.0051	1.0003
Morrisons	1.0005	0.9942	0.9997	0.9987	1.0020	1.0005	0.9990
Sainsbury's	0.9904	1.0121	1.0013	1.0088	0.9942	1.0028	0.9997
Tesco	0.9992	0.9987	1.0071	1.0010	0.9965	0.9990	0.9985
Waitrose	0.9811	1.1000	1.0061	1.0124	0.9843	1.0023	1.0068
Iceland	0.9997	0.9982	1.0058	0.9975	0.9991	1.0001	1.0021

Source: Newing *et al.* (2015)

Table 7.9 Observed vs predicted model fits in Cornwall.

52 week average – 2010 trading year	*Status*	*Ratio of observed to predicted store revenue*
Store 1	Calibration store	0.99
Store 2	Calibration store	1.00
Store 3	Calibration store	0.97
Store 4	Calibration store	0.98
Store 5	Control store from collaborating retailer	0.96

Source: Newing *et al.* (2015)

we would expect high-income consumers to be willing to travel further to visit higher end retailers, and low-income consumers to exhibit a willingness to travel further to reach a discount retailer, notwithstanding the fact that β has been set to increase the impedance of distance for low-income consumers. This version of the model was used by Newing (2013) and Newing *et al.* (2015) to produce a working model for a major UK supermarket retailer in the region of Cornwall in the UK. The results were extremely impressive. Table 7.9 shows the observed versus predicted revenues at a selection of stores based on this model (see also further discussion in Chapter 11).

7.4.2 Retail brand and customer loyalty

The representation of brand in the models is becoming ever more important as more and more emphasis is being placed on brand identities by retailers to help maintain and win market share. A particular issue in spatial modelling is the question of inertia imposed by customer loyalty to specific brands, which can be observed in many markets. In the automotive market, for example, the amount of loyal custom has a marked effect on customer patterns. An early realisation of this problem came with our observation that car manufacturers with no dealer representation on the Isle of Wight in the UK (a small island economy linked to the mainline by sea only) did not lack market share, which the models found hard to 'understand'.

The brand loyalty effect is also strongly observed in the retail travel market. A major international travel agent (let's call it TravelShop) was considering a major rationalisation programme in its branch network in response to pressure from the internet. The directors determined that closures should focus on retail centres with more than one outlet, of which there were more than 200 in the UK. They wished also to understand better the likely impact of outlet performance on retail business and market performance. For many years, TravelShop had operated a 'gravity model' for store performance, with store attractiveness represented through a combination of floor space, a regional brand weighting and pitch within a centre measured by pedestrian footfall.

To respond to their directors, the market analysis team went back to analyse some historic data for 12 outlets, which had previously been closed, in multi-outlet centres. What the team found was that the loss of customers through this process was systematically under-predicted by the existing gravity model. The team also found that a great deal of variability existed (actual variations from 10 per cent to 79 per cent; model variations from 16 per cent to 31 per cent). The team turned to us for advice on how to improve the model. We argued that these findings were not surprising if (local) markets were considered to be *saturated*. Here the definition of saturation is that there are so many outlets that customers can determine their preferred retailer on the basis of brand characteristics (e.g. product, advertising, previous experiences).

If a market is saturated, then the market share is purely related to the presence or absence of a retail brand. Thus (following the argument of Birkin *et al.* 2010),

$$\theta_j^k = \delta_j^k \omega_J^k \Big/ \sum_k \delta_j^k \omega_J^k \tag{7.5}$$

where δ_j^k is 1 if brand k is present in centre j, *0* otherwise; $\omega_J{}^k$ is a brand weighting for competitor k in region J ($j \in J$).

In a more saturated market, if one TravelShop branch closes while others remain open, its sales are simply transferred to a neighbouring outlet. In contrast, if a local market is perfectly unsaturated, then the market share is a function of individual outlet attractiveness. Hence,

$$\theta_j^k = \sum_{m \in k} \theta_j^{km} = \sum\nolimits_{m \in k} W_j^{km} \Big/ \sum\nolimits_{km} W_j^{km} \tag{7.6}$$

In this case, the different retail organisations (k) can open new outlets and attract a share of the market in proportion to the attractiveness of an outlet, which is related to floor space (F_j^m), pedestrian footfall (G_j^m) and regional brand attractiveness:

$$W_j^{km} = \omega_J^k F_j^m G_j^m \tag{7.7}$$

The corollary is that in perfectly unsaturated markets, when an outlet is closed, lost sales are transferred between the remaining outlets in proportion to their attractiveness.

In reality, local markets are unlikely to be completely saturated, and equally unlikely to be perfectly unsaturated. Therefore we introduce s, a *saturation parameter*; where s=1 denotes that the market is completely saturated, and

s = 0 denotes that the market is completely unsaturated. Aggregate market share is simply a weighted combination of the two models (7.5) and (7.6):

$$\theta_j^k = s\left\{\delta_j^k \omega_J^k \Big/ \sum_k \delta_j^k \omega_J^k\right\} + (1-s)\left\{\sum_{m\in k} W_j^{km} \Big/ \sum_{km} W_j^{km}\right\} \tag{7.8}$$

(7.5) (7.6)

We estimated a value for the saturation parameter using customer retention data from the 12 sample stores (s=0.23). Using the revised model, we calculated retained business at between 27 per cent and 68 per cent. The average error in predictions of the value of business retained in each centre was reduced from 26 per cent to 15 per cent. This degree of variation is still less than was observed in practice, but a great improvement on the fair shares model. A number of factors remained outside the control of the model, such as migration from outlets to virtual channels, and the opening or closing of competing outlets. Micro-location is another factor that we found to be important. as retention levels were much higher between branches that are physically close together. Many of these factors were included in later versions of the TravelShop model, which was used to plan a major phase of consolidation in the TravelShop retail network.

Another example can be drawn from a scenario for a financial service provider wishing to open an additional branch in a large retail centre where it is already present with a number of existing branches. The inadequacies of a simple SIM used in this context can be demonstrated by comparing results with what we might intuitively expect. All things being equal, we would expect total centre sales to remain unchanged or undergo only a marginal increase. We would expect sales (for a new branch) to be generated from the centre's existing branches – deflecting more heavily (depending on brand loyalties and competitor mix) from branches of the same brand. What we actually see from the standard model is an increase in centre sales, near-incremental sales for the new outlet and minimal deflections from the existing same brand branches. However, the model must be fine-tuned to deal with many combinations of changes. For instance, we could equally envisage a scenario where a retailer may wish to close one of a number of branches in a smaller provincial or rural centre. A standard model formulation typically results in a large migration of business to surrounding centres and little or no business retention for the remaining branch. A more realistic outcome would be for a small number of customers to migrate to banks in other centres and for the retained branch to pick up some of the lost sales. For a financial service client, such a refined model allowed corporate retention levels of 18–25 per cent when branches closed, compared to a standard entropy model that gave company retention rates of only 3–5 per cent. The same kind of brand loyalty and network connectivity issues are evident whether we are modelling at the centre or individual outlet level.

7.4.3 Network interdependence

We have thus far argued that many retail networks are influenced by loyalty to brands. However, in some situations, the reverse can be found where performance indicators such as average sales per branch are found to increase with levels of representation. The argument for this is that organisations need multiple 'touchpoints' in order to provide acceptable service to their customers. Financial services organisations are a good example, where customers expect to be able to process counter transactions at different locations; for example at work, from home, while shopping, or at multiple points in between. Figure 7.4 is based on the performance of a major UK retail bank with over 1,000 outlets throughout the country that are not uniformly distributed. It is a scatterplot of their share of all bank branches in each of 51 regions in the country versus their market share of business in the same regions. Intuitively one might expect a linear relationship between branch share and market share but we can clearly see that branch share increases as a non-linear function of market share. With the bank, we used these results to identify where investment in new branches would generate the biggest growth in share. At some point branch saturation will set in and new branches would begin to cannibalise revenues. (Also see the interesting approach to corporate networks suggested by Williams and Kim 1990.)

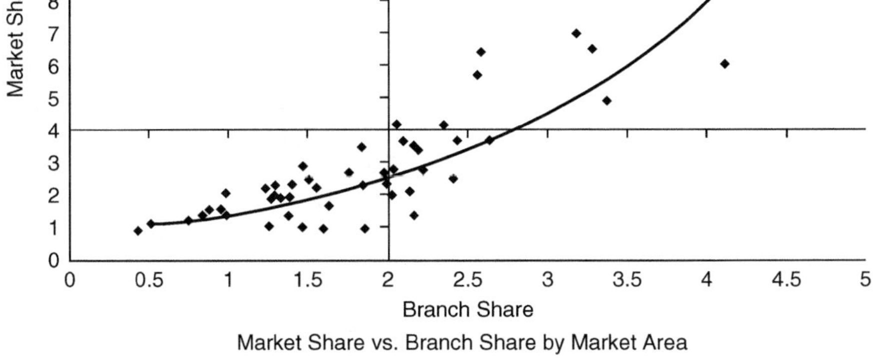

Figure 7.4 The importance of the network effect in retail modelling.
Source: Authors

A relatively simple way to incorporate the network effect is to introduce a branch share parameter that can form an extra component of the compound attractiveness model described earlier, that is,

$$W_j^{Nm} = \sum_j \delta_{Ij} F_j^m \Big/ \sum_{jm} \delta_{Ij} F_j^m \tag{7.9}$$

where δ_{Ij} is 1 if branch *j* is in region I, but zero otherwise.

7.4.4 Hierarchical destination choice

The work of Stewart Fotheringham in the 1980s was especially pertinent to the issue of considering the modelling of flows to either stand-alone and/or clustered shops (Fotheringham 1983, 1986). He argued that the standard SIM should be adapted in situations of this type, where the alternative destinations are far from independent (and thus in violation of the 'independence of irrelevant alternatives' assumption that underpins the entropy-maximising approach). His reformulated model includes a *competing destinations term* that recognises that outlets or centres in very close proximity to each other are really a single destination in the eyes of consumers. In a standard entropy SIM, if four outlets of the same size and attractiveness are equidistance from an origin zone, the likelihood that one outlet would be visited over the others would be 25 per cent. However, if three of these four outlets are close to each other then, in reality, consumers may perceive that there are really just two destinations and each of the three clustered outlets would be more attractive than the isolated one (when discussing this with students, a good example to use is bars or public houses).

To capture this 'hierarchical' destination choice, Fotheringham (1983, 1986, 1988) introduced another term to the model – an accessibility score which measured how close each store was to its neighbour (cf. Hansen 1959). The higher that score the more attractive the store would become. The appropriateness of this new term depends largely on the type of good (comparison or convenience) and whether one is talking about single stores or stores within centres. However there is plenty of empirical evidence to show the model can improve revenue estimates in a number of cases (see for example retail and non-retail examples of the model in Ishikawa 1987; Pellegrini *et al.* 1997; Fotheringham *et al.* 2001; Hu and Pooler 2002; de Mello-Sampayo 2009).

7.5 Conclusions

This chapter has examined the nature of store or centre attractiveness in detail. We observed in the introduction that 'attractiveness' of a retail centre is typically given as *Wj* in a spatial interaction model. We then reviewed a plethora of studies which have tested and refined the attractiveness term. The major question, therefore, is which should be used in any particular application? There is a clear issue here of art versus science to model variable selection. We have discussed the science above – the art comes from the user's expertise in what is most appropriate in which circumstances.

For those readers keen on mathematics, we end by summarising a version of the model which takes many of our disaggregations above into account. In fact, this was the highly disaggregated version of the model we used in many applications in the car market for example, where the data is extremely rich:

$$W_J^{kln} = \omega_J^{kln} \sum_{j \in J} R_j^{kl} Z_j^k \gamma_{\Psi(j)}^{\varepsilon(j)kl\text{Æ}(j)n} \mu_j^{kl} \text{Ç}_j^{kl} \tag{7.10}$$

where

ω_J^{kln} is a measure of centre performance
R_j^{kl} is store performance
Z_j^k is retail floor space
$\gamma_{\Psi(j)}^{\varepsilon(j)kl\varphi(j)n}$ represents brand loyalty
μ_j^{kl} is store agglomeration
Ç_j^{kl} reflects the maturity of an outlet.

This version of the model is hard to implement in many markets because of the lack of data. Thus the choice of which variables to use probably becomes a mixture of pragmatism (what is possible in the time allowed) and necessity (what really must be included).

8 Network optimisation

8.1 Introduction

In this chapter we explore the concept of optimisation in retail location planning. We present three very different examples of how optimisation might be applied to support location planning. The first relates to the design and implementation of techniques for finding optimal sales territories. This is appropriate when retail firms demarcate sales territories to individual stores or outlets in order to try to prevent intrabrand competition by ensuring each individual outlet has a reasonable degree of autonomy within its sales territory (although of course cross-boundary flows across such areas cannot be prevented in their entirety). In some countries and in some markets this is also a legal necessity in order to offer protection for incumbent retailers from further competition from their own (often franchised) organisation. These types of optimisation models are also frequently used in logistics to allocate territories to each distribution depot in a retail firms' distribution network (see Fernie 2009 and Fernie and Sparks 2009 for more details here). Perhaps this is with the aim of minimising overall total distance travelled between depots and outlets (see also the discussion on network analysis and GIS in Chapter 3).

The second type of optimisation involves finding 'best' locations for stores given certain operating criteria or constraints (i.e. every region must have a minimum sales level or market share). As we have seen in the book so far, the majority of store location tasks involve evaluating the impacts of actions relating to individual outlets or small groups – that is, openings, closures, refurbishments, relocations etc. However, from time to time, retailers may be interested in asking questions such as 'what would be the optimal set of stores for my company in region X?' This may reflect a long-term goal of moving an entire network of stores from a set of current locations {X}, perhaps built up haphazardly over many years, to a new set of locations {Y}, based on a more up-to-date spatial analysis of the contemporary market. It would also be a useful question to ask for a retailer moving into a new region. For example, given the short-term goal of opening ten new outlets from scratch, where would be the set or combination of locations {X} that would maximise sales in that new region?

The third example explores the usefulness of a type of optimisation procedure for mergers and acquisitions. To date we have said little about this growth strategy. However, retail mergers and acquisitions produce unique problems – especially, how to deal with two store networks once the companies merge or get taken over and become one organisation. Too often the response of the newly created, larger organisation is to opt for rationalisation, closing outlets where there is any spatial overlap in individual towns or cities. We would argue that such a strategy may result in very profitable stores being closed simply because they are in close proximity to an existing outlet. Instead, we offer a more flexible post-acquisition network optimisation strategy.

We shall now explore all three types of optimisation problem below.

8.2 Optimisation territory allocation: customer marketing areas

In this section we explore the construction of 'customer marketing areas' (CMAs) in the car dealer market, especially using the case study of work undertaken in the past for the Ford Motor Company. In many areas customers may wish to purchase a car from a local dealer near to where they live, but to have the car serviced close to where they work, or shop. However, this could mean buying a car from a Ford dealer, but having it serviced by another dealer representing a different marque. This kind of competition for business is normal for the dealers, but damaging to the Ford brand.

Another important effect is that many European markets are now thought to have too many dealers with very low or even negative profitability. Ford has therefore utilised the CMA concept as a means of identifying larger dealer territories in a structured way. This has allowed the Ford brand to be projected more consistently across a stronger network of dealers. Although the number of dealerships may have been reduced, this is not necessarily true of customer access points, since dealers have been encouraged to diversify their networks within a single CMA. Thus, aftersales support may now be provided through dedicated 'service-only' satellite points, at more convenient and distributed locations than the main dealer sales points. In addition, the concept of the CMA allows retailers to explore a new regional level of performance. Our experience was that retailers seemingly had two ways to look at their market performance. The first was an obsession with national market share. The second was a focus on individual store performance. What we introduced was a third approach that sat between these two levels: a regional or CMA perspective. The argument is that retail performance and planning should additionally take place at the CMA level.

The creation of CMAs has some similarity to the derivation of 'Functional Regions', first developed at the Centre for Urban and Regional Development

Studies (CURDS) at Newcastle in the early 1980s (e.g. Coombes *et al.* 1986). The CMA approach is, however, distinctive with respect to other regionalisation approaches in at least two respects:

1 It is a general-purpose approach that has been applied in a number of European countries. For example, Table 8.1 shows a pan-European regionalization, in which 19 countries have been split into self-contained regions on the basis of minimal cross-boundary interactions between the various regions.
2 It forms a basis for commercial applications.

In practice, CMAs can be generated using a variety of data sources, which often includes retail interactions in addition to journey-to-work flows and data for specific businesses. For example, in defining a set of CMAs for a bank or building society, it would be appropriate to include information about account ownership, transaction behaviour, and ATM usage by the organisation's own customers.

Putting all this information together we can identify regions of the country where the vast majority of individuals live, work and shop – these are what we define as CMAs. Figure 8.1 shows one of these CMAs for Cambridge in the

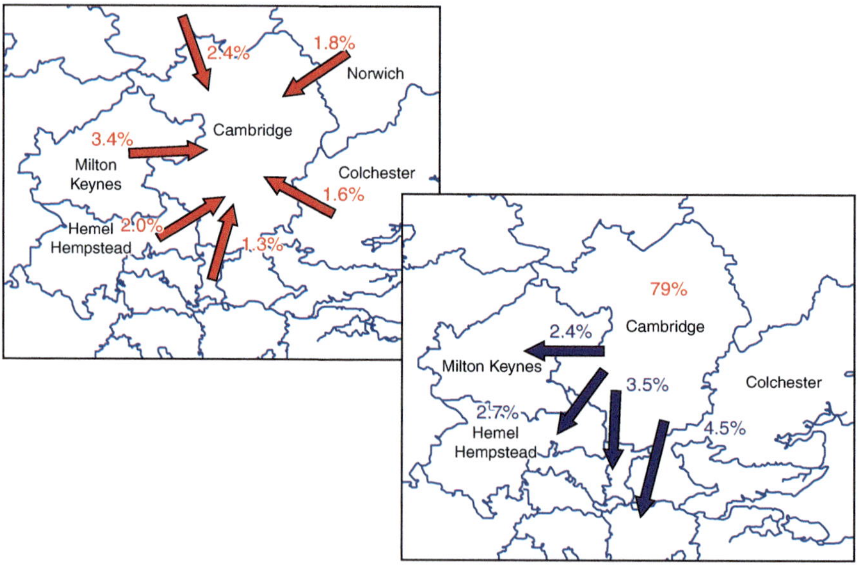

Figure 8.1 Journey-to-work flows in and out of Cambridge CMA in the UK.
Source: Authors

Table 8.1 European containment area solutions.

Country	*CMAs*	*Population*	*Area (Sq Km)*	*Population density (Pop/Sq Km)*	*Average population per CMA*	*Average JTW containment*
Austria	24	7795786	83641	93	324824	77.6%
Belgium	29	9941896	30513	326	342824	74.5%
Denmark	20	5213472	42805	122	260674	80.3%
Finland	16	5098754	331261	15	318672	88.1%
France	124	56738473	547135	104	457568	82.4%
Germany	149	81262432	355646	228	545385	82.9%
Ireland	26	3525719	68496	51	135605	n/a
Italy	119	57062670	302293	189	479518	85.9%
Netherlands	43	15344668	34872	440	356853	80.1%
Norway	19	4344482	316054	14	228657	93.9%
Portugal	25	9868037	88834	111	394721	82.6%
Spain	83	39433942	505575	78	475108	82.1%
Sweden	26	8816381	438937	20	339092	86.8%
Switzerland	32	6888037	40826	169	215251	86.4%
UK	110	54853618	243907	225	498669	80.1%
Poland	49	38127514	308711	124	778113	n/a
Hungary	20	10362129	92887	112	518106	85.2%
Czech Rep	23	10321345	78251	132	448754	92.2%
Greece	18	10313687	131505	78	572983	

Note: CMA-Customer Marketing Area

east of England. The figures show that relatively few people who live in this CMA work or shop outside the CMA, and similarly not many people outside the CMA shop or work inside it. Figure 8.2 uses the CMA concept to compare the performance of a UK retailer in different CMAs. Using six different performance indicators we can map out these spider web graphs and quickly compare performance in different geographical areas.

Figure 8.3 displays a CMA map for a car manufacturer in Spain. Most Western countries have the requisite information to generate such CMAs. Again, key performance indictors can be plotted for each CMA, as also shown in Figure 8.3.

Birkin, Clarke and Clarke (2002: 185) cite the appraisal of the strategic application of containment areas for Ford motor company in a newspaper report in 1998:

> “In this report, Ian McAllister, Chief Executive of Ford United Kingdom, explains how the concept of the Customer Marketing Area (CMA) has been applied across the whole of the UK dealer network. In part, the rationale for this has been to eliminate destructive ‘Intrabrand

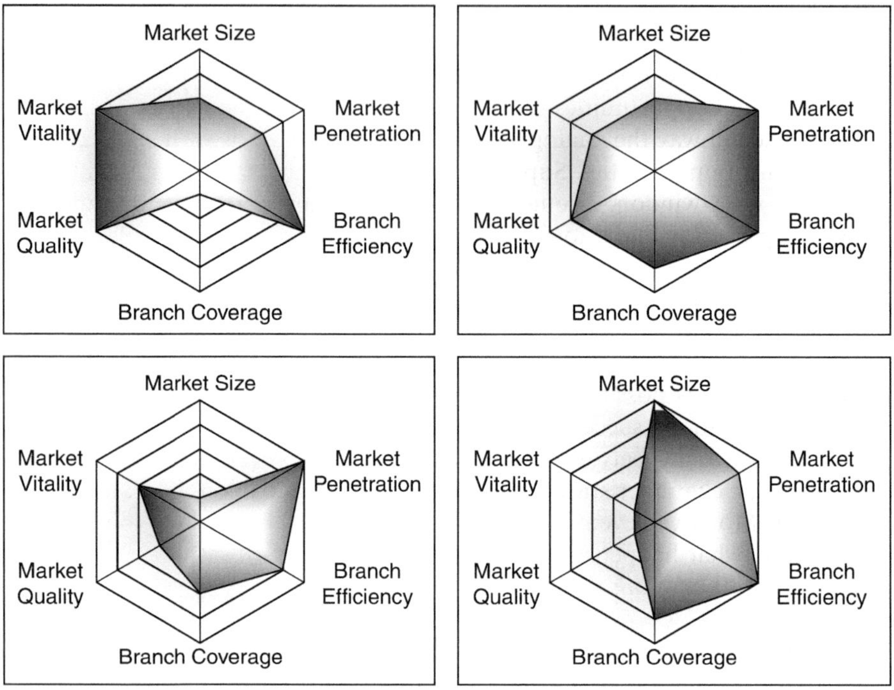

Figure 8.2 Analysing variation by CMA: benchmarking performance and opportunity.
Source: Authors

> competition' between dealers. Intrabrand competition arises from the proliferation of dealerships within relatively small geographical areas, and can lead to fragmentation of the brand."

8.3 Retail store network optimisation

8.3.1 Background

As noted above, most retail networks in the UK and elsewhere are essentially 'legacy networks', in that they have evolved over a long period of time. Much of the expansion of a particular retailer's network has probably been through organic growth, opening a number of new stores each year. In addition, some retailers may have also grown part of their networks through mergers and acquisitions. As a consequence of different strategies undertaken at different times most retailers' networks are not optimal. That is, if they had to start again with a blank sheet of paper they would not create the network they currently operate. A good example of this is presented in the following simple

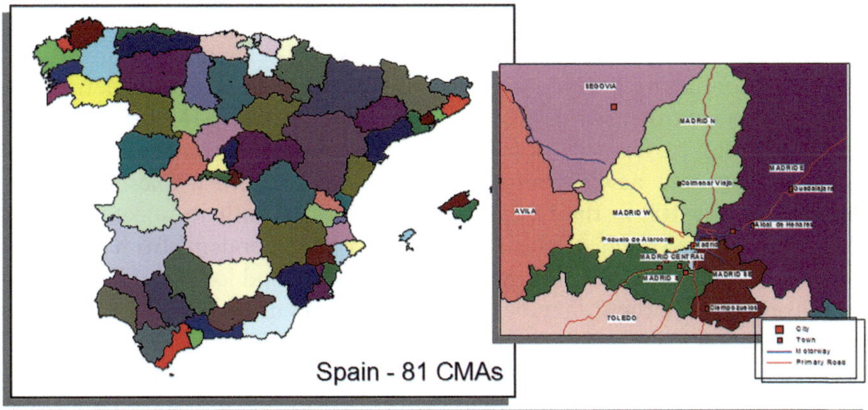

CMA No.	Customer Marketing Area Name	Region Type	Containment Levels (%) JTW	Customer NCS	Dealer NCS	Projected 2002 Industry Volume All Segments Retail	Fleet	MCV	LCV	Total	Population	Area (sq km)	Popn Density
013	AVILA	Rural	n/a	80.1	75.7	2,454	8	227	330	3,020	176,312	8,281	21
016	SEGOVIA	Rural	n/a	92.2	90.9	2,756	1	246	277	3,280	146,579	6,966	21
017	MADRID W	Metro	59.4	47.6	54.7	18,550	12,542	914	985	35,320	785,964	1,668	471
018	MADRID CENTRAL	Metro	38.7	50.4	57.0	18,786	12,701	1,546	1,221	36,424	725,808	71	10,251
019	MADRID SE	Metro	37.6	61.4	56.2	21,570	14,584	2,472	2,469	41,381	907,416	861	1,053
020	MADRID S	Metro	58.3	73.5	59.6	19,832	726	1,912	1,777	24,247	949,103	1,380	688
021	MADRID N	Metro	49.0	37.3	42.7	23,456	15,859	1,621	1,680	41,627	702,178	2,159	325
022	MADRID E	Metro	55.5	35.9	50.1	27,608	18,666	1,781	2,009	46,268	1,055,001	13,304	79
061	TOLEDO	Town	94.4	82.0	81.0	9,399	51	1,230	1,230	11,909	546,649	16,347	33

Figure 8.3 Building CMAs in Spain for a major car manufacturer.
Source: Authors

example. The authors worked with a major British clearing bank which had 1,996 branches (early 2000s). However, many of these branches had greatly overlapping catchment areas and it was felt that fewer, better positioned, outlets could offer the same level of geographical coverage. Figure 8.4, using network optimisation techniques described in this chapter below, demonstrates that a much reduced network of just 296 locations (plotted on the map) would ensure that 80 per cent of the UK population would still be within 15 minutes' drive-time of an outlet of this bank.

Our second, more detailed, case study relates again to the car market. At first sight, the car industry does not appear to be the most natural application area for location planning techniques. Surely, motor vehicle sales are determined by testosterone, branding, lifestyle, safety and panache? In fact, relative location is as important as any of these factors, and more important than most. Consider the case of Renault, for example, which in the UK is a middle ranking, mid-market marque. In the top 100 towns where a Renault dealer is located, the manufacturer enjoys a 10 per cent market share. In the top 100 towns lacking a Renault franchise, the share falls below 5 per cent.

Suppose a new car dealership is planned in a city or region. The likely sales of the new dealership may be estimated by considering the number of customers in its environs, their buying preferences and behaviour, and the distribution of competing automotive franchises that would seek to penetrate the same market. In other words, a spatial interaction model (SIM) will provide a highly appropriate framework for estimating dealer performance.

Now suppose a new manufacturer wishes to create a whole network of dealerships, say 100 in the UK. Not only is it important to select 100 good locations but also each of these should be 'interdependent' from the other locations. Therefore, even though there may be 50 good locations in London, only five or six of these may be required, whereas in Leeds only two of ten or twenty good locations may be needed. The locations must be assessed, not only individually but also in relation to the coverage and competition between each another. This is the idea behind the Idealised Representation Plan (IRP) developed by the authors to enable retailers to explore optimisation more fully. The idea can be expressed in non-mathematical terms as:

> *Find a network of N outlets so that this network sells more cars than any other network of N outlets, assuming that sales within the network may be predicted using an appropriate spatial interaction model.*

It is also possible to generate versions of the IRP in which the number of outlets to be found is also a component of the problem: for example, to find the best network of outlets to deliver a certain market share or sales target. However, these alternative formulations are, in essence, a special case of the specification provided above. Thus, to find a network that delivers 10 per cent market share, one would probably begin by estimating the number of dealers required, then running the IRP. According to results, the estimate of the number of dealers required would be updated and the IRP run again, and so on, until a satisfactory solution is reached. Note also that it is possible to specify the IRP so that an optimum level of profitability is achieved. In principle, this is straightforward and looks attractive, but, in practice, this requires that costs be estimated at each location, which can be difficult. Furthermore, in the car market, dealerships are typically franchised, as we have already seen. In this case, assuming that car prices are relatively fixed, as they generally are, then manufacturers really do want to maximise their sales and the cost of doing business is a concern for the franchisee.

Following our description above, the IRP may be viewed as an embedded SIM. In essence, the trick is to run a SIM millions of times with different dealer locations, in such a way that the best solution can be found. The IRP, therefore, has the same data requirements as a conventional SIM. In the case of the car market, the availability of appropriate data is typically very good. In many countries, vehicle licensing agencies, such as the Society of Motor Manufacturers and Traders (UK) and AAA (France) and others, collect vehicle registrations at small-area level by marque and model. Thus, in any of

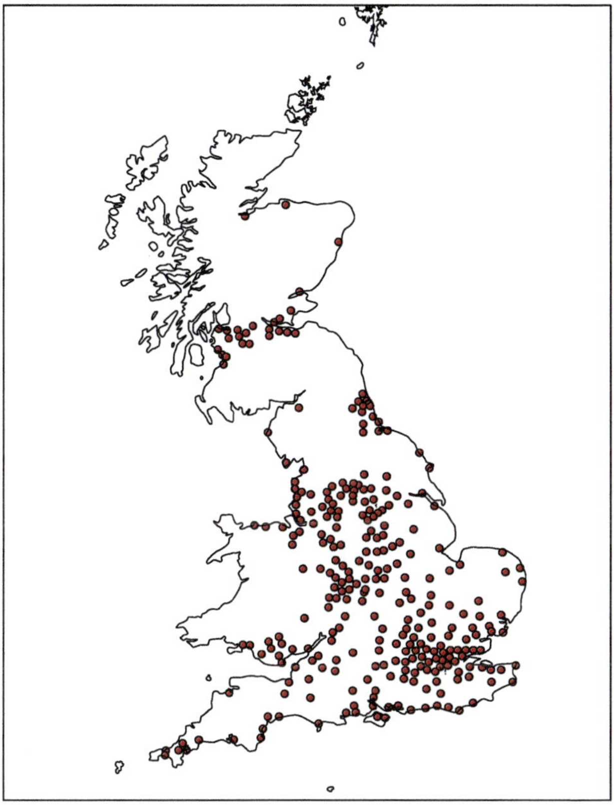

Figure 8.4 The 'optimal' locations for branches of a major UK clearing bank.
Source: Authors

these countries, one can find the number of Ford Mondeos, against VW Golf, Audi A4 or any of a variety of competing models, at postal sector or equivalent small-area level. When combined with dealer lists, which can typically be obtained from individual manufacturers, an excellent platform for modelling is provided.

The structure and mathematics of the IRP heuristic are described in full by Birkin, M. Clarke and George (1995). Imagine that an ideal network of 100 dealers is to be found. Initially, one would find the single best location for a dealer. This is done by imagining a car dealer at every possible location (e.g. every postal sector or demand zone) and running the associated SIM to calculate dealer sales. The location that generates the highest sales is selected as the first location. This dealer now becomes 'fixed', and the procedure is repeated to find the next best location. The first dealer is then 'perturbed' (moved around) to check that it is still in the best place. This procedure is repeated until 100 locations have been identified.

An example of an IRP for the country of Denmark is shown in Figure 8.5. The objective was to take an existing network of dealers and to optimise the network, so that the same number of cars could be sold through fewer outlets by finding the ideal locations. This process is illustrated for just one manufacturer, a major manufacturer with more than 100 dealers in the existing network. The left-hand map of Figure 8.5 shows the existing network of branches, while the right-hand map shows the re-configurated network with fewer dealers, but designed to sell as many cars as the larger network on the left-hand map.

An objection that is sometimes raised against IRP is that, usually, it is impossible to adopt a carte blanche approach to planning. For example, if established retailers want to find the best locations for their 200 stores, then surely they are somewhat constrained by the investment that has already been made in the network. This argument is less strong in the automotive sector for a number of reasons. First, as we have just seen, distribution is franchised, with the manufacturers generally having most power. Therefore, manufacturers are relatively free to massage their networks if they feel it is appropriate. Second, many manufacturers are genuinely moving into new geographical markets. In particular, established brands such as Ford have begun to open up

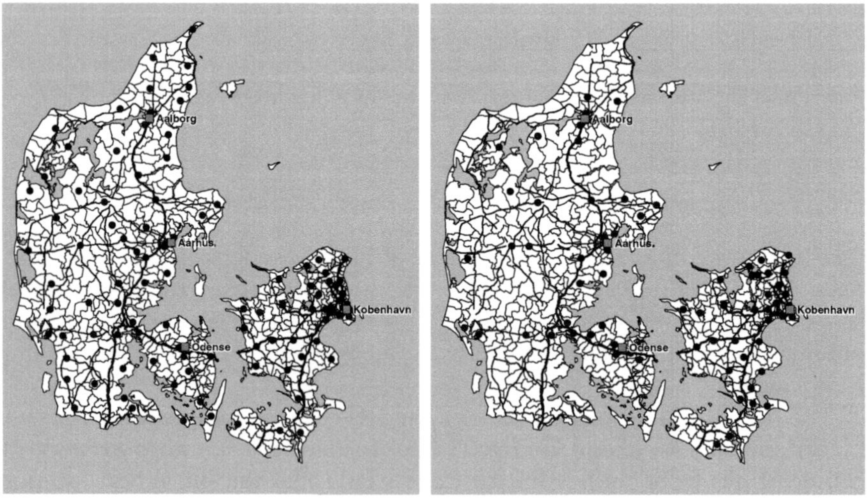

Figure 8.5 The results of the Denmark IRP: left map shows actual distribution of dealers; right map shows the optimal distribution.

Source: Birkin *et al.* (2002)

new markets in Eastern Europe and, to a lesser extent, Africa and the Middle East in recent years. Relatively new brands, such as Kia and Proton, have also come into play and clearly demand new networks, whereas the introduction of new products and branding strategies, such as electric cars and Smart vehicles, provides an impetus for innovative thinking. Finally, the changing nature of both demand and supply in the automotive sector has led to significant changes in market planning, for example, larger dealer areas of responsibility and product specialisation.

Other problems with the IRP heuristic described above are the quality of the solutions and the practical deployment of the outputs. The IRP heuristic is unlikely to produce poor solutions to the network optimisation problem. Equally, however, it is well known that hill climbing procedures such as this will rarely yield perfect solutions to complex problems. This has been demonstrated for sample IRPs by Birkin *et al.* (1995). George *et al.* (1997) have described an alternative solution procedure that uses a genetic algorithm to find improved solutions. The genetic algorithm does not work with a single solution, as with the conventional IRP heuristic; rather, it works with a population of different solutions which are refined and combined in such a way that progressively better solutions are allowed to 'evolve'. For more details, see George *et al.* (1997), Gen and Cheng (2000) and Konak *et al.* (2006).

It is difficult to overemphasise that the IRP is an extremely hard problem in both computational and theoretical terms. The classic Travelling Salesman Problem is a famously difficult computational problem when the salesperson needs to visit more than about 20 places in the correct order. The IRP has many similar features, but, typically, involves not only the choice between many thousands of locations, but also the interactions between individual pairs of thousands of locations. However, the effort necessary to overcome these issues is well spent, because of the importance of this class of business planning problems.

If IRPs are to prove useful to planners, they must clearly be capable of deployment. Whether the network to be managed is already established or completely new, it will be necessary to manage the plan as a living and dynamic entity, and not as a lifeless and static blueprint. For example, an 'ideal location' cannot be commandeered – good sites, which may or may not be close to the ideal, will need to be acquired when they become available. Also, as noted above, retailers may wish to find only the optimal locations for new sites, keeping their existing networks largely intact. A powerful feature of the IRP is that it is able to incorporate existing network constraints – for example, if we want to create a new network of 100 dealers from a combination of 50 existing dealers and 50 new ones. This is an important consideration for practical applications. Birkin *et al.* (1996) showed the results of an optimisation procedure for Toyota in Seattle, USA. The task assigned to the IRP was simply to find two new 'optimal' locations in Seattle that would minimise the impact on existing dealers. Thus all new sales would largely come from the company's competitors. If this type of optimisation procedure could be replicated across all US towns and cities the benefits to the company would be huge. Taking very aggregate statistics, if 1,000 new sales can be generated at an average of $20,000 and an average profit margin of 15 per cent, then net profit could be as high as 1,000 * 20,000 * 0.15 (=$3 million) *per region.*

8.4 Optimisation following mergers and acquisitions

Mergers and acquisitions have always been an important retail growth strategy. Many organisations have used this strategy not only to increase corporate power by eliminating competition, but also to gain access to markets where they have been previously under-represented. Such activity can produce rapid geographical growth although there are risks and heavy costs associated with both strategies. Burt and Limmack (2001) note that between 1982 and 1996 there were over 1,000 retail takeovers involving British companies alone. Wrigley (2000) provides an interesting table of major merger and acquisition activity in the global retail industry in the mid to late 1990s. Some famous UK examples include Boots merger with Alliance Unichem in 2005 (the new AllianceBoots itself bought by KKR, the private equity company in 2007); Morrisons' acquisition of Safeway (UK) in 2004 and Carphone Warehouse and Dixons in 2014. In the USA many of the major retailers have expanded aggressively through mergers and acquisitions. Albertson's became the second largest grocery firm in the USA when it bought Safeway (US) for $8.2 billion in 2014. Of course some mergers or alliances take place on a truly global scale. For example, Ford acquired Jaguar in 1990 only to sell Jaguar to the Indian giant corporation Tata in 2008. British Airways merged with Iberia to form the third largest air carrier in the world in 2010. BP and Arco announced merger plans in August 1998, creating an organisation with a turnover of over £67 billion (although again Arco was largely sold off to Tesoro in 2012). In these and other cases the quest is to guarantee economies of scale through reduced overheads, greater economies of scale in buying and selling and better product development

The merger/acquisition planning process is typically described as having three main stages – the pre-acquisition planning stage, the negotiation and bidding stage, and the implementation and integration stage. The pre-acquisition stage is concerned with evaluating the best partner for merger within the context of the overall corporate goals and objectives of a company. It is perhaps the most essential element of the acquisition process as it serves as the foundation for the process that follows (Hubbard 1999). It is also one of the most overlooked areas of the process, and as a result, one of the main causes of poor merger performance (Hubbard 1999). Although retailers may search for take-over targets that are a good business or geographical 'fit', in practice these choices may be constrained by non-spatial considerations – such as price, availability, ownership structures, product compatibility, management structures or simply the personal preferences of senior executives. Many of these factors become more important as retailers seek quick and defensive strategies of growth (Laulajainen 1988). Nevertheless, we argue that the spatial dimensions and, more specifically, the potential performance of branch networks are critical factors. That is, in the retail sector, unlike in other sectors, the performance of a merger (whether a merger achieves its pre-defined objectives in terms of, say, market share or sales targets) is ultimately achieved through the aggregate performance of all the individual outlets acquired. Therefore, the performance (or potential performance) of the distribution network is an important, (if not the most important) criterion in evaluating best partners for merger.

As we noted above, prior to merger or acquisition there is often little consideration of the ideal partner in terms of access to new geographical markets. There are notable exceptions. The Leeds brewer and public house operator Tetleys used mergers and acquisitions strategically to undertake a process of contagious diffusion from Leeds, obtaining significant post-war market shares in Sheffield and Bradford through acquisition before merging with Walkers of Warrington to obtain high market share in Lancashire and Cheshire. Further mergers with Ansells saw them diffuse to the Midlands before acquiring national presence through an amalgamation with Allied Lyons. Another example is provided by the battle for Scotland in the grocery market. The two leading UK grocers, Sainsbury's and Tesco had little presence in Scotland by the end of the 1980s. As they perceived saturation to be more imminent in their homelands, Scotland became a fierce battleground. Tesco eventually won the battle in 1994 when they purchased the Scottish grocery retailing chain Wm Low & Co. plc. (see Sparks 1996).

A third example is that of Carrefour in France. Burt (1986) tells how in the 1970s and early 1980s they cleverly bought up parts of chains in different regions of France to increase their national coverage (which they could eventually turn into full acquisitions as desired: the partial buyouts helped them in part to circumnavigate Loi Royer introduced in 1973 to try to prevent new superstore development in France). Then, in 1991 they bought out the troubled Montlaur company to give them greater presence in the southwest of France. This was followed by the acquisition of Euromarche in the same year, producing what Burt describes as 'a perfect geographical fit' (Burt 1986: 157).

Finally there was the merger of BP and Arco in 2000. On agreeing the merger in 1998, BP's Chairman emphasised how BP's high market share of petrol sales in eastern USA would be complemented by Arco's high market share in western USA: 'For one thing it brings "coast to coast" coverage in the USA refining and petrol retailing market, combining Arco's West Coast strengths with BP Amoco's east of the Rocky Mountains' (quoted in Fagan 1999).

Having identified a suitable target, the negotiation and bidding stage is concerned with the technicalities of the transaction, that is, securing the acquisition of the target at the right price. It involves developing a bidding strategy (which is effectively a 'battle' plan detailing the various lines of attack, outflanking manoeuvres, and counter attacks), evaluating the value of the target and negotiating the price, obtaining the necessary finance and, finally, closing the deal. This stage is predominantly external to the acquirer and target organisations – being carried out by legal advisers and financiers (for more information see Sudarsanam 1995).

The implementation and integration stage is concerned with the practicalities of bringing the two organisations together after the deal is complete. It involves a re-evaluation of organisational and cultural fit issues – given that at this stage more information is available than at the pre-acquisition stage, the functional integration of the various elements of the two companies involved (IT systems, personnel, distribution networks etc.), and ongoing monitoring of the performance of the new or combined entity (again, for more information see Sudarsanam 1995). Once two branch networks are brought together by such circumstances, there is likely to be considerable overlap in their spatial representation, especially if these are national rather than regional players. The merger of the Leeds Permanent and the Halifax Building Societies in 1995 is a case in point, with 78 per cent of the original Leeds Permanent outlets residing in the same financial centres as a Halifax branch. In such cases, the gut reaction of the new combined organisation is often to undertake closures per se – that is to remove the overlap where a small town now has two branches of the same organisation. Similarly in 2014 Carphone Warehouse and Dixons reported a merger in the UK worth around £3.5 billion. The corporate lawyer dealing with the case was quoted as saying, 'I expect there will be substantial rationalisation over time, maybe not immediately.' 'From past experience, watching similar mergers in recent times, there has been greater rationalisation than expected' (both quotes in This is Money.co.uk 2014).

Similarly, in 2013 in the USA, a spokesperson for Office Depot reported that following their merger with OfficeMax, 400 stores would be closed in order to remove the overlap between outlets, saving $75 million.

> 'One of our 2014 critical priorities is to improve our store footprint in North America,' Chief Executive Officer Roland Smith said in the statement. 'The overlapping retail footprint resulting from the merger provides us with a unique opportunity to consolidate.'
>
> (quoted in *Bloomberg News* 2014)

Many other organisations have reported the same practice. For example, both Barclays and Lloyds TSB stated their intentions to take this course of action in their planned mergers with Woolwich and Abbey National respectively. Barclays announced their plan in 1999 to close all Woolwich outlets within 100 metres of their existing outlets, while Lloyds TSB announced in 2001 its plan to close all Abbey National overlapping outlets within a quarter of a mile of their own (although subsequently this merger was blocked by the UK Competition Commission).

Such strategies, however, may not be the best way to proceed, producing little flexibility in very different geographical markets. Instead, we believe merger activity produces tremendous opportunities for the new organisations to review and revise their network distributions. Flexible, local solutions should be central to that review process. The merger/acquisition activity offers the retailer the opportunity to make a detailed, localised examination of the performance of both networks. In fact, two stores in the same town might both be performing extremely well – closing one simply because it is nearby might be a false economy. To help build a post-merger/acquisition strategy we suggest a different form of optimisation. In this case the procedure works by running the location models (SIMs in our case) for every town or shopping centre for eight different scenarios. The scenarios are shown in Figure 8.6.

To illustrate this concept, we draw upon a real example – the combined Barclays/Woolwich networks following their merger in 2000/01. More details appear in Birkin, Clarke and Douglas (2002), which the following text is based on. The new network contained 2,292 branches and generated a market share of 14.5 per cent in the UK retail current, financial services savings, and mortgage account markets in 2000. For ease of analysis, we focus on a subset of this network in the Kent CMA in south-east England. In this CMA, the combined network totalled 67 branches and achieved a 17 per cent market share of the three product markets, some 2.5 per cent above its national average. The locations of each branch in the network and the distribution of market share at this time are shown in Figure 8.7.

Network Location

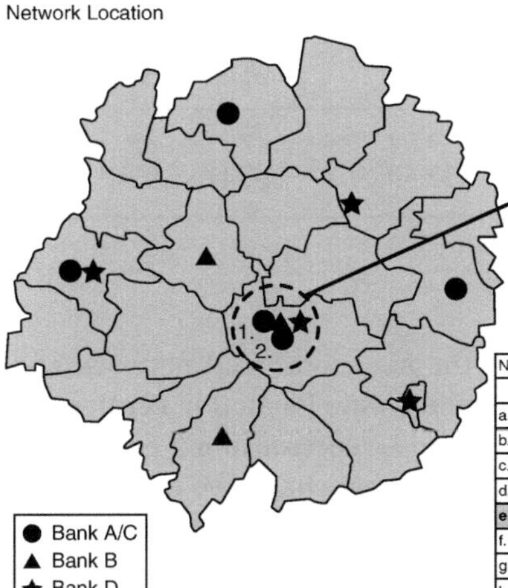

Network Configuration Strategies

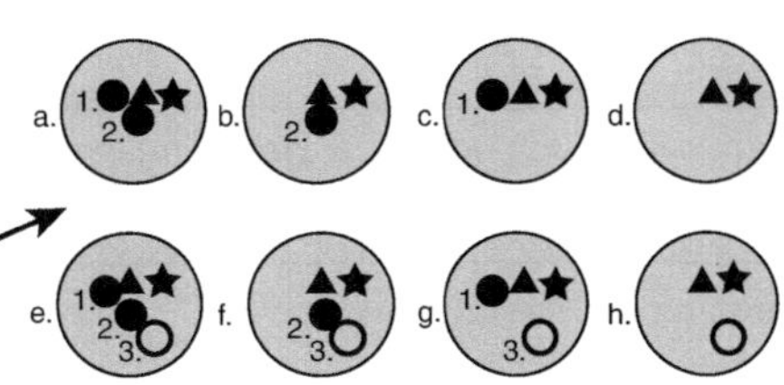

Network Strategy	Mkt Share (%)	Revenue (£million)	Costs (£million)	Profits (£million)
a. Retain Existing	15	0.4	0.2	0.2
b. Close 1. Retain 2	7.5	0.2	0.1	0.1
c. Close 2. Retain 1	7.5	0.2	0.1	0.1
d. Close Both	0	0	0	0
e. Open New (3) + Retain 1&2	**22.5**	**0.6**	**0.3**	**0.3**
f. Open 3. Close 1. Retain 2	15	0.4	0.2	0.2
g. Open 3. Close 2. Retain 1	15	0.4	0.2	0.2
h. Open 3. Close 1&2	7.5	0.2	0.1	0.1

Figure 8.6 An optimisation model to evaluate different network configuration strategies.

Source: Birkin *et al.* (2002)

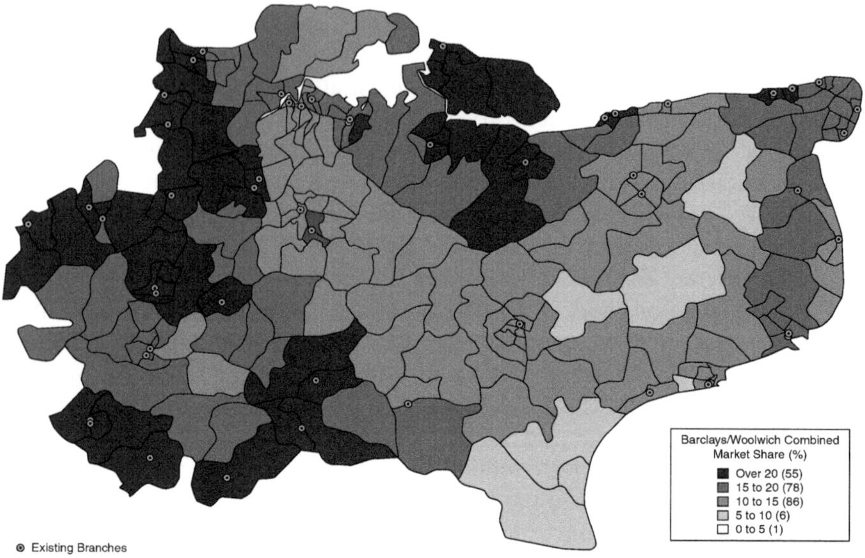

Figure 8.7 Location of branches and market share following Barclays/Woolwich merger (2000).

Source: Birkin *et al.* (2002)

Table 8.2 Key indicators for the Barclays alternative strategies.

Strategy	*Outlets*	*Sales*	*Market share (%)*	*Sales/outlet*	*% change*	
					Volume	*Sales/outlet*
Retain existing	67	48 845.75	17.03	729.04	na	na
Barclays strategy	57	41 809.39	14.58	733.50	−14.41	0.61

Source: Birkin *et al.* (2002)

In many retail centres throughout Kent the new combined network contained more than one outlet. In fact, 24 branches were located in centres that contained one or more outlets of the same brand, representing a 36 per cent overlap in the CMA as a whole. In terms of market share, we also see that there were considerable variations throughout Kent. For example, in the west, market share reached 35 per cent in some postal sectors, whereas in the south-east it fell to less than 5 per cent.

As noted above, Barclays' planned strategy was to retain an existing Barclays outlet but close any Woolwich branch within 100 metres of an existing Barclays branch. On the basis of this simple criteria, ten outlets were earmarked to close in Kent. Table 8.2 compares key indicators for the original

67 branch network compared with the same indicators for the new, revised 57 outlet strategy. As we can see, the strategy is estimated to reduce new account generation by 7,000 accounts (sales here refers to new account generation per annum) even though there would be an estimated marginal improvement in network efficiency (by 0.61 per cent – measured in terms of sales per outlet). Market share for the new combined organisation was estimated to decrease from 17 to 14.6 per cent because of a 14 per cent decrease in revenue volumes (versus a strategy of retaining the existing distribution). As outlined above, this estimated decline in sales and market share mirrors evidence from other recent UK retail financial mergers. In fact, given the merger objectives of maximising market penetration (measured in terms of revenues and market share), the strategy achieves the opposite of what Barclays intended. Again, the question is, therefore, what would have been the optimum configuration of the combined network? Could they have done better? Let us now, therefore, perform a search for the optimum combined network configuration.

As outlined in Figure 8.6, there are a number of network configuration options that we can consider in a search for the 'optimum'. Table 8.3 shows the results of implementing the full optimisation model shown in Figure 8.6.

Table 8.3 New indicators for alternative strategies for Barclays.

Strategy	*No.*	*Network size*	*Sales*	*Sales/outlet*	*% change*	
					Volume	*Sales/outlet*
Retain existing	na	67	48 845.75	729.04	na	na
Barclays strategy	na	57	41 809.39	733.50	−14.41	0.61
rationalisation and relocation	1	60	53 764.90	896.08	10.07	22.91
	2	59	52 823.76	895.32	8.14	22.81
	3	58	52 506.93	905.29	7.50	24.18
	4	57	52 784.20	926.04	8.06	27.02
	5	56	49 979.55	892.49	2.32	22.42
	6	55	51 018.89	927.62	4.45	27.24
	7	54	51 599.74	955.55	5.64	31.07
	8	53	50 274.10	948.57	2.92	30.11
	9	52	49 736.44	956.47	1.82	31.20
	10	51	49 822.73	976.92	2.00	34.00
	11	50	49 647.36	992.95	1.64	36.20
	12	49	49 165.15	1003.37	0.65	37.63
	13	48	48 355.77	1007.41	−1.00	38.18
	14	47	47 599.74	1012.76	−2.55	38.92
	15	46	47 345.55	1029.25	−3.07	41.18
	16	45	46 755.06	1039.00	−4.28	42.52
	17	44	46 404.22	1054.64	−5.00	44.66
	18	43	45 435.34	1056.64	−6.98	44.94
	19	42	45 485.60	1082.99	−6.88	48.55
	20	41	44 632.58	1088.60	−8.63	49.32

Source: Birkin *et al.* (2002)

So, when running the eight scenarios it is possible to close, open and relocate branches of either Woolwich or Barclays. Such a localised mixed approach yields some interesting results.

If we assume that Barclays would have preferred strategies which would rationalise the network overall, we can see that it would have still been possible to achieve both cost savings and efficiency gains, while at the same time, increasing revenues. We find that the best configuration – again based on maximising revenues and efficiency (equally weighted) – is probably strategy 7. This strategy, in which 27 of the existing branches are retained, 27 are relocated, and 13 are closed, improves network efficiency (and subsequently network profitability) by 30 per cent. In addition, the strategy also achieves a 5 per cent improvement in revenue volumes, compared with the existing configuration. This is a 30 per cent improvement on Barclays' preferred integration strategy! The distribution of this new network configuration is shown in Figure 8.8. It can be seen that in some centres, where the combined network results in a situation in which there are now two outlets within one hundred metres of each other, both are retained. In fact, only four of the closures are the same as those under Barclays preferred strategy.

8.5 Conclusions

In this chapter, we have looked at the issue of optimisation and introduced the concept of three different types of optimisation procedures which might be useful to retailers at different stages of their development. In particular, we have introduced the concept of the Idealised Representation Plan and have discussed examples from the automotive and financial service sectors. We note, however, that typically these problems are both mathematically and computationally complex, so that obtaining effective solutions requires a blend of analytical imagination and computational brute force. However, given the ready availability of ever-increasing quantities of computing power, the ability to implement procedures of this type can only get easier. Hence, the central message of the chapter is that extra computational power can provide the means to solve new and interesting classes of locational problems.

We would also argue that it is important that store location teams investigate the impacts of mergers/acquisitions on local revenues and market shares for another key reason. Increasingly such activity is coming under scrutiny by national and more global competition commissions (names vary around the world). These agencies are increasingly concerned with excess monopoly power and will investigate more and more as power becomes concentrated in to the hands of fewer and fewer retailers. Even if a merger or acquisition is allowed, these agencies may demand outlets are sold in geographical regions where monopoly power may be deemed to be too concentrated. When Asda took over Netto in the UK for example in 2010, it had to sell 47 out of 193 stores in areas where the UK Competition Commission felt they would have too much of a local monopoly on trade. (For a longer discussion see

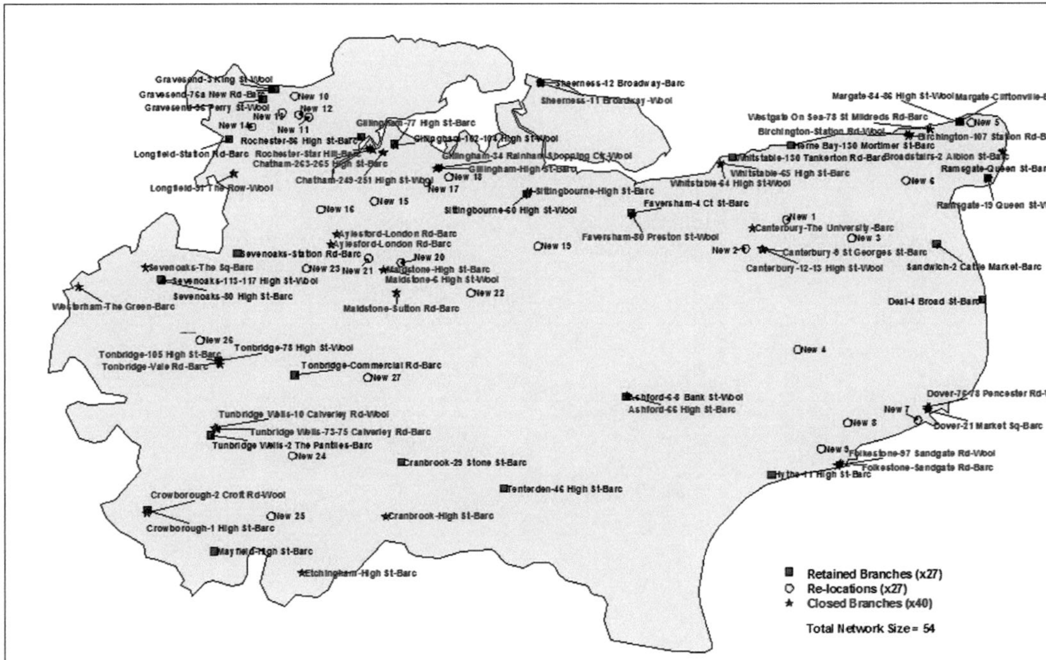

Figure 8.8 New configuration of stores for strategy 7 in Table 8.3.

Source: Birkin *et al.* (2002)

Competition Commission 2011, and also analyses of different market share scenarios for potential mergers in the UK grocery market: for example, Poole *et al.* 2002a; Hollingsworth 2004; Hughes *et al.* 2009.) It is clear that geographers can play a major role by looking at the impacts of potential merger activity on market shares and the likelihood of concerns from organisations such as the various global competition commissions using optimisation techniques to develop a competition commission-friendly future strategy.

9 Network reinvention

In this chapter, consideration will be given to ways in which entire retail networks can be overhauled or 'reinvented' in the context of rapidly changing service environments. The first part of the chapter reviews factors that have given rise to 'retail turbulence' in recent years. This recaps on some issues that have been raised in earlier chapters but it is important to reiterate key changes as context for the material in this chapter. Second, it will be suggested that one of the consequences of turbulence has been to increase the complexity of retail processes. The argument will be advanced that one of the consequences of complex and turbulent markets is that conventional modelling approaches may no longer be necessarily the best available techniques for the analysis task. Some supplementary and alternative methods will thus be introduced, including branch scorecards, network segmentation, representation planning and investment appraisal. Finally, a number of case studies will be considered in which a combination of spatial analysis techniques have been used to support network reinvention.

9.1 Recap on retail turbulence

The UK baby boom generation of the 1960s have lived through a huge shift in retail markets during their adult lives. Some of the major factors which have driven change within retail markets are illustrated in Figure 9.1. Since many of these individual factors are interconnected and overlapping, they will be combined here through loose groupings of underlying trends:

9.1.1 Changes in demand

There can be little doubt that *customer power* has increased in recent years (see also Chapter 6 for further examples). Of course certain consumers are as a whole more affluent than ever before, for example, according to *Trading Economics* (2014) gross domestic product for the UK at constant prices doubled between 1980 and 2010. But they are also more mobile – for example, cars on the road have risen from 19 million in 1971 to over 35 million in 2013 (Department of Transport 2014) – and better informed through the

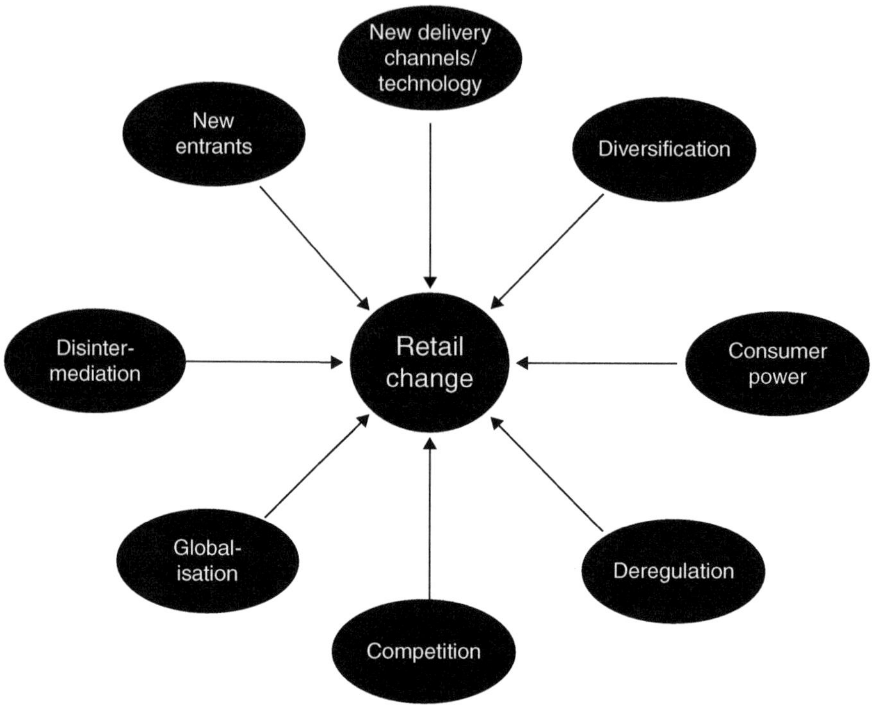

Figure 9.1 Retail market turbulence.
Source: Authors

rapid growth in communication technologies. One of the most obvious consequences of this trend is the increasing adoption of 24/7 retailing. Thirty years ago very few retailers were open on Sundays, and almost none on a 24-hour schedule. Now many supermarkets, as well as local stores, can be patronised on every day of the week, right through the day and night. Tesco alone, for example, had 394 outlets open around the clock in 2010 – twice as many as the number of police stations offering similar service levels according to the *Telegraph* (2010). An increasingly diverse range of formats, and in particular a rapid expansion of 'convenience' stores has been another important response to this pressure (discussed more fully in Chapter 2).

Ever more complex, multipurpose trip-making is another factor here which has obvious importance for spatial modelling, as it is likely that retailing becomes a secondary, even a distress activity associated with a primary focus on travel to work, school, a leisure activity or social focus. Another manifestation of this trend has been towards niche spatial marketing, where retailers are able to customise their range and formats according to the demographic profiles of specific local markets (see Chapter 2 for more details). An example is UK food retailer Sainsbury's introduction a few years ago of ready segmented oranges for the lunchtime market in some of its city centre outlets. These trends have also been backed up by an ability to target customers ever more closely as retailers are becoming less dependent on crude market research classifications based on social class or demographics, but are more likely to exploit data captured from loyalty cards, or at least individual-level data from lifestyle surveys for this purpose. More details on these trends can be found in Chapter 2.

9.1.2 Changes in provision

Whereas 30 years ago international retail brands were largely confined to the motor vehicles and petrol retailing sectors, there is now a well-documented trend towards *globalisation* throughout retailing (e.g. Coe and Wrigley 2009). In the supermarket sector, for example, Tesco has been active especially in Eastern Europe and the Far East; and while major entrants into the UK market include Aldi and Lidl, among the Big Four, Asda itself is now of course only a small part of the world's largest retail corporation, Walmart. On the high street more generally, the 'clone city' phenomenon (Glendinning and Page 1999; NEF 2005, 2010) is perhaps clichéd but evidence is now manifest in cities all over the world of international brands like Starbucks, McDonald's, Subway, Zara, French Connection and many others. Here globalisation has gone hand-in-hand with *deregulation* as not only have domestic markets become porous to foreign competition, but the exclusive monopoly position of individual retailers for certain products has been broken. A simple example would be that 30 years ago only Post Offices were able to sell postage stamps in the UK, but now this product is available through an extensive range of outlets. In return, the Post Office itself has been able to compete in new markets such as the provision of financial products, a move that has also been manifest among the supermarkets: for example Sainsbury's Bank. An even more significant trend of retail *diversification* has been unleashed, as UK supermarket retailers, for example, have also moved increasingly into sectors such as clothing and stationery; but have found their own core markets under threat from organisations such as high street discount chains selling a mixture of household goods and packaged groceries.

9.1.3 Changes in technology

Here perhaps the changes are most profound of all. Thirty years ago, cash could only be withdrawn from a bank branch in major UK cities between the hours of 9.30 a.m. and 3.30 p.m. five days a week. Access to cash, as well as other services such as account balance enquiries, has been radically transformed for a long time by the introduction of cash points (or ATMs) as *new delivery channels*. Now of course account enquiries and alerts are easily supported by mobile telephones which were not even invented 30 years ago, and likewise the internet, which has affected financial services organisations in a much more profound way.

As well as the obvious effect of supporting web-based delivery through companies such as Ocado, these new technologies have allowed the *disintermediation* of retail markets, allowing manufacturers to supply directly to customers, although in practice this has been facilitated by the introduction of new intermediaries ('reintermediation'). Among the most significant of these new intermediaries are the price comparison sites such as Kelkoo and GoCompare, increasing price awareness in particular (alongside diversification as discussed already), which has served to increase levels of *competition* in all retail markets. This process has been reinforced by the ability of the internet in particular to support *new entrants* to retail markets often with quite modest infrastructure requirements and start-up costs. Of course Amazon is the most spectacular example of such a business which has sprung up into a retail giant from nothing over the last 20 years.

One vital consequence of these many retail changes has been the increased concentration of retail networks. For example, since the mid-1980s the number of UK bank branches, car dealers, petrol stations and newsagents (CTNs) have all been reduced by a factor in the order of 50 per cent or more (see Figure 9.2). These numbers are thought-provoking in themselves but of

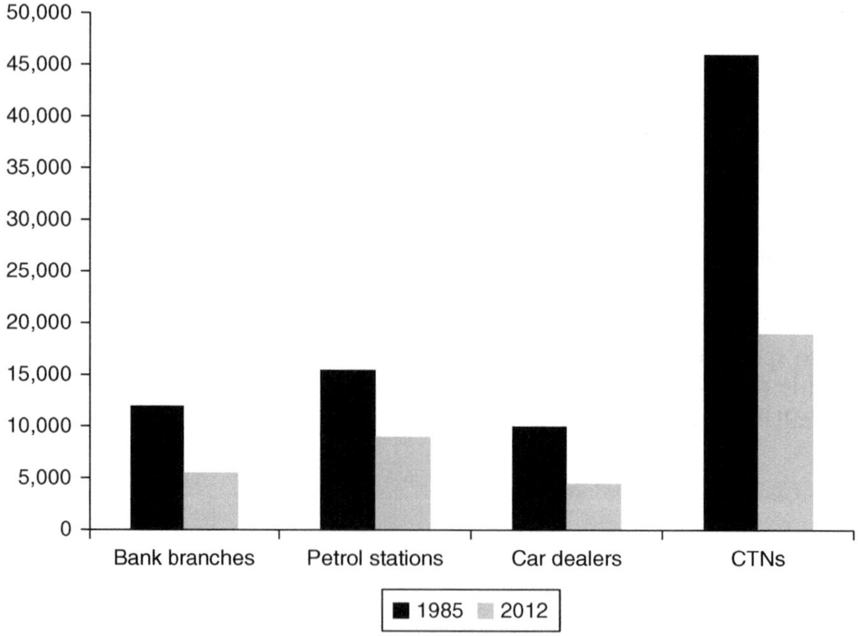

Figure 9.2 Changes in UK network densities over time.
Source: Authors

course this only tells half the story in light of our previous discussion, as the range of products on sale, the way they are promoted and the means for their distribution have all changed dramatically. In short, therefore, retailers have faced major challenges in the reconfiguration of their networks over an extended period of time; and, for the current purpose, the way in which they have used spatial modelling to assist in this process is the topic of major interest. In the third part of this chapter we will therefore consider a number of examples of the use of spatial analysis technologies among major retail corporations to support business transformation. One of the key ingredients in this process, for example, is the spatial interaction model (SIM) which is discussed extensively elsewhere in this volume. However, other techniques have also been important, and it is to these ancillary techniques that we now turn before the case studies are considered.

9.2 Methods to support network reinvention

Elsewhere in this volume (see Chapter 5 in particular, although examples appear throughout the book), the benefits of SIM as an aid to retail location planning are pressed strongly, and of course the virtues of this approach are many. However, in this section we consider some of its limitations, and argue that in some cases alternative techniques are required. This will be done with reference to a specific range of issues, many of which overlap with the discussion of the previous section in this chapter; and also in relation to a variety of retailers across different sectors. In this way we hope to make the point that market turbulence necessitates that retailers of all types need to use a variety of spatial analysis techniques to support business transformation, including but by no means limited to SIMs. This will be followed by a consideration of some alternative techniques, and all of this then paves the way for the case studies which follow in Section 9.3.

For the purpose of this discussion, let us assume we begin with a completely standard specification of a production-constrained SIM (cf. Chapter 5):

$$S_{ij} = O_i^{W_j e^{-\beta c ij}} \Big/ \Sigma_j W_j e^{-\beta c ij} \qquad (9.1)$$

in which O_i is a population count in small area i; W_j is floor space provision in retail centre j; c_{ij} is the straight-line distance from i to j; and S_{ij} is the number of people living in i and shopping in j. Beta is a parameter to be determined. On the horizontal axis of Table 9.1 a variety of named organisations from different retail sectors are identified. The essence of the problem here is to consider how such a basic SIM needs to be shaped and extended to meet different requirements. The circles in the table are intended to show examples where a particular factor is thought to manifest itself quite strongly in a specific retail context. These examples will be used to substantiate the discussion which follows, although in practice an equally interesting exercise for the reader might be to work through the table box-by-box and grade the importance of each factor, say on a scale of 1 to 10. The point is that many of the factors are significant, to a greater or a lesser extent, across many of the sectors.

For convenience of organisation, and also to make the connections to the previous section of this chapter as clear as possible, we will arrange the column headings of Table 9.1 into the three categories adopted previously, that is, consumption, provision and technology.

9.2.1 Consumption factors

A logical starting point is to consider the impact of *demographics* as a factor in itself. Few retailers will draw customers from anything like a full cross-section of the population. Some, such as Mothercare, will necessarily pull from restricted and well-defined groups. In the examples here McDonald's can be expected to rely particularly heavily on younger customers (from a wide spectrum of backgrounds) while Starbucks would most likely draw from a young but more affluent customer base, who have the inclination to socialise over coffee and the means to disregard the financial impact (again more details appear in Chapter 6).

The significance of *work-based trips* has already been noted (see Chapter 6). In earlier publications we have described work with UK book retailer WHSmith (WHS), whose city centre sites have always been among the most prosperous (Birkin and Foulger 1992). For example, Holborn in London is one of the busiest locations and the flow of commuter traffic is a clear reason for this. Indeed WHS is a particularly interesting case having expanded rapidly as a seller of newspapers in railway stations in the middle of the nineteenth century (WHSmith 2014). The expansion of other retail outlets on railway station forecourts, most notably in recent years Marks and Spencer (who have also extended this strategy to motorway service stations) is an eloquent testimony to the increasing importance of work-based retail activity. Another excellent example, shown in Table 9.1, is petrol retailing where this level of dependence is probably at its highest, for obvious reasons. Similarly, banks and other financial organisations have a long history of providing services to customers with access from work as well as home.

Similar trends are manifest in relation to *leisure trips*. Indeed airport shopping in particular has been a major growth sector for retail floor space in recent years (Thompson 2007). Cinemas are another leisure attractor in their own right, and the up swell of multi-screen cinemas at the expense of the smaller independents has typically meant a move to brownfield or out-of-town locations with retail adjacencies (especially quick service restaurants). Other leisure attractors might include football grounds, Natural Trust properties, and maybe even pubs and clubs – although whether these latter should more properly be considered as retail adjacencies is a moot point. Indeed both 'work' and 'leisure' should be understood here as very wide-ranging categories, thus absorbing activities such as school and university ('work') as well as

Table 9.1 Complications for models of retail interaction.

Organisation	Retail type	Work-based trips	Leisure trips	Elastic demand	Low-value purchases	Internet competition	Alternative channels	Brand	Price	Quality	Demo-graphics	Scale economies	Footfall and micro-location
Tesco	Supermarket								●	●			
Shell	Petrol station	●	●										
Post Office	Convenience				●								●
Ford	Motor vehicles							●	●				
Barclays	Financial services	●					●						
McDonald's	Quick service restaurant			●							●		
Vue	Cinema		●	●									
Thomas Cook	Travel agent					●							●
Starbucks	Coffee shop			●							●		
Waterstones	High street retail					●						●	●
Boots	Convenience				●								
Argos	Variety						●						
Marks & Spencer	Department store							●	●	●		●	
Next	Clothing							●	●	●			

Source: Authors

social trips or church attendance ('leisure'). Other activities, such as visits to the doctor or hospitalisation are less easy to categorise, but the same lessons hold true.

Low-value purchases are important because any choice modelling process assumes that the decision about where (and when) to shop involves a conscious evaluation and trade-off between different aspects, such as accessibility and product range. If a lot of money is involved, for example in buying a new car or a piece of furniture, then this assumption is reasonable. If the product is something like a newspaper or chocolate bar then it is (very) much more likely that a decision is impulsive and linked to other considerations, such as work or leisure activities as discussed above. Another issue here is that low-value, isolated purchases are much less likely to be tracked, for example through loyalty schemes. It is no coincidence that the motor industry has enjoyed the benefit of excellent customer (interaction) data for years (e.g. Birkin *et al.* 1996) whereas Post Offices and convenience stores still have negligible information about their walk-in customers. This lack of data is significant in one of the case studies in Section 9.3.

9.2.2 Provision factors

Among the traditional 'four Ps of marketing' model, alongside our primary concern with Place, the other questions of Product, Price and Promotion also need to be considered in any retail model (Birkin *et al.* 2010).

The *quality* (of the *product*) is a vital consideration for many retailers, and thus for example it is hard to compare Tesco with Waitrose, Zara with Next, or McDonald's with Carluccios, even though the markets they serve are ostensibly similar (i.e. groceries, clothing and restaurants respectively). It is worth noting that some products tend more towards homogeneity that others, and it is in relation to some of the more homogeneous products that internet sales have grown most rapidly in recent years, for example, travel tickets, books, CDs and now downloads (see also discussion at 9.2.3 and Chapter 10). It is equally important to recognise that the delivery environment may have a less tangible importance for the delivery of such products – for instance, book buyers may prefer to browse within a large retailer in preference to an internet site, even though the final purchase itself is identical (see also Weltevreden 2007, who highlights the potential contribution of alternative channels in the purchase of different products).

In the imagination of the customer, the quality of the product is intimately linked to the *brand* (and its *promotion*). Hence, a supermarket retailer such as Sainsbury's in the UK might be able to accommodate a higher price even for a completely uniform product such as a named brand of ketchup or baked beans than a competitor such as Morrisons because of the way in which these retail brands have been developed over a long period. An important related issue is customer loyalty, which applies not just to high-value products such as motor vehicles but across the whole range of high street retailers, supermarkets, petrol stations, financial services providers etc. (Grewal *et al.* 2004). The crucial point for the current discussion is that retail behaviour cannot be viewed as a unique decision process but has to be related to previous decisions and purchases.

The importance of price variations has generally been underplayed within retail modelling practice. The fundamental economic principles are clear and unambiguous here – higher outlet prices means limited attractiveness and reduced retail spending. Spatial price variations were absorbed into location theory more than half a century ago (e.g. Moses 1958) although in general this work draws more heavily on the inspiration of the classics such as Alfred Weber's Theory of the Location of Industries (Weber 1909) in emphasising production rather than the distribution of commodities. Perhaps it is the fact that the interrelationships of price, quality and brand are so intricate that makes this problem more difficult than at first appears. For example, it is clear that price makes a vitally important contribution to the selection of products in the motor vehicles industry, and yet the price of a similar marque such as a family saloon will vary radically between manufacturers such as Skoda and BMW. Even petrol retailing – perhaps the most homogeneous of markets in this list – shows substantial price variation at quite a local level according to perceptions of brand and the kinds of loyalty scheme which predate most others in the business as well as the principles of spatial clustering (Heppenstall *et al.* 2006).

Another fundamental of economic production which also has importance for the process of retail distribution is the existence of *scale economies* (or 'agglomeration'). In this sector the most important manifestation of scale economies is in comparison retailing, where multiple competitors actually benefit from co-location since customers benefit from the ability to select from an extended range of product alternatives. Thus an agglomeration of clothing retailers on a city centre high street will be to the common good relative to a single unit in a more peripheral location. In the major department stores this idea is extended under a single roof as not only alternative brands but increasingly competing franchises are represented in close proximity (for example, House of Fraser in Leeds has concessions for retailers like Austin Reed and Monsoon, even though those fascias also have independent representation in the city centre). One of the interesting developments within the last few years has been an extension, which is perhaps surprising, into the motor vehicles sector through the introduction of 'multi-franchising'. Thirty years ago, exclusivity was universal, even to the extent that international regulations on retail competition were warped to protect the special relationship between dealers and manufacturers (the 'block exemption' – Brenkers and Verboven 2006). Nowadays it is far from uncommon to see multiple brands clustered at the same location, often within a single retail unit. This is another example of growing consumer power, although also indicative of how growth in the dealer groups has begun to challenge the hegemony of the manufacturers themselves.

Spatial aggregation, specifically the assumption – sometimes rather casual – that individual retailers can be loosely grouped into 'centres' can also be challenged through the recognition of tremendous variations in *footfall and micro-location* factors. In PhD research conducted with the support of the market analysis consultancy GMAP Ltd, Eyre (1999) made an assessment of many micro-location factors important to the performance of a high street retailer, ranging from parking and pedestrianisation to the frontage and layout of individual stores (see Chapter 7 for more details). In many sectors, the consolidation of brands through a process of mergers and acquisitions has emphasised multiple representation within a single centre and thus helped to expose variations between stores that are similar in all respects other than their location and format. Notable examples in recent years include many mergers in the financial services sectors, for example, Barclays and Woolwich Building Society; Lloyds and TSB; Halifax and Leeds Permanent, and later Bank of Scotland. Thus in the retail sector generally, both Morrisons' acquisition of Safeway and the combination of Thomas Cook and Going Places are both good examples; although due to the sheer extent of its branch network the Post Office remains the retailer with the greatest localised performance variations, as we shall see again later.

9.2.3 Technology factors

The advances in computational technologies have begun to transform the retail process over the last decade, and this transformational process must ultimately be recognised in any spatial analysis of the marketplace. *Internet competition* has introduced a whole range of new organisations into retail markets. Some of these, (e.g. Amazon) are operating head-to-head against established players. Other new businesses are changing the market itself in more radical ways. For example, in the travel sector companies like Expedia and TravelBag are able to exploit the introduction of new product delivery mechanisms such as e-tickets to weaken the hold of established players quite dramatically. The markets for books and travel are often quoted as examples of sectors in which the impact of e-commerce has been particularly transformational (see Farag *et al.* 2006; also Chapter 10); nevertheless these effects have been felt in all sectors and will continue to become more and more important, in particular with the emerging popularity of m-commerce and interactive television.

One should note that internet competition also has effects that are explicitly spatial – notwithstanding variations in the uptake of technology, access to internet retailing is almost universal and not strongly conditioned by physical accessibility as in conventional store-based retailing. Thus Weltevreden (2007) has found that once introduced to the idea of internet retail, customers in rural areas are more frequent repeat buyers than their urban counterparts. While this conclusion is drawn from limited data at a relatively early stage in the e-retail adoption cycle (it draws on panel data from 2002 in which only a few thousand customers had used any electronic channel) it seems a safe bet that these trends will continue and become more strongly embedded. We consider some evidence of our own in Chapter 10.

Established retailers are far from completely impotent in the face of changing technology, and in particular debates over 'bricks and clicks' (Gulati and Garino 2000) serve to emphasise the adoption of *alternative channels* for product delivery. Thus the supermarket retailers seem to have largely fought off the challenge of usurpers like Ocado and Webvan by leveraging their extensive store networks and mighty logistical capabilities by simply assimilating internet delivery capabilities alongside their core business, albeit with varying levels of sophistication (see Sparks and Burt 2003, for further discussion). Furthermore, alternative channels are not just a substitute for traditional mechanisms. Farag *et al.* (2006) have also shown the need for complementarity between different retail channels and emphasised the importance of multi-channel retailing to businesses in all sectors. For example, a consumer might wish to compare items on the high street before placing an order at lower cost over the internet, but retailers working effectively across both channels would be those best placed to succeed. This mix looks set to become even more convoluted with the advent of a new generation of technologies including retailing by mobile telephone (m-commerce) and interactive television. All of these trends are considered in more detail in Chapters 10 and 11.

To summarise this rather detailed argument, simple SIMs (or even 'gravity models') may have had some traction in the 1960s and beyond when considering major strategic decisions such as the impact of a new shopping centre like Haydock or the Metro Centre in the UK (Guy 2006). For a more precise evaluation of the fortunes of individual retailers, including an assessment of issues such as local profitability, competitiveness and cannibalisation of trade more sophistication is required; and as markets have become more complicated (as we argue here) this therefore becomes more difficult.

Of course an obvious response to many of these issues is to refine the models themselves, and indeed there could be many gains to be made from this approach. Certainly the models can be disaggregated to allow for much more differentiation in the purchasing patterns and behaviours across demographic groups (see Chapter 5 and Birkin *et al.* 2002, 2010 and Newing *et al.* 2015) and more subtle representations of micro-location and localised clustering may be introduced (Birkin *et al.* 2010). Various market segmentations can be employed to stratify consumer groupings (Birkin *et al.* 2010) and ultimately this could lead to powerful agent-based representations of customers, although progress to date has tended to focus rather more on the retailers themselves (Heppenstall *et al.* 2006). Agglomeration economies and hierarchical decision-making processes can be approximated through either extending the range of components to store attractiveness (Eyre 1999) or through refinements such as the Competing Destinations model (Fotheringham 1983: see also Chapter 7 for more details).

However, some of the problems introduced by complex trip-making and the lack of definition in low-value transactions are an order of magnitude less tractable again. According to Birkin *et al.* (2003), some activities are simply rather hard to model. Petrol retailing is a prime example which is more weakly determined by residential customer origins, and strongly linked to convenience passing trade or on occasions 'distress purchasing'. Other categories could be low-value sales categories such as newspapers and confectionery, and convenience or impulse activities such as cash withdrawals.

In the next section of this chapter, we therefore propose a simplified range of network evaluation methods which suggest themselves as appropriate to these difficult situations. On the one hand, regression models are easily available within desktop spreadsheets and statistics packages and can be applied to an enormous range of problems for the estimation of relationships. At the other extreme the technique of branch scorecard modelling has been developed as a focused approach to the problem of network reinvention. First, it's useful to briefly repeat the difficulties with existing approaches (see Chapter 5 for more detail).

A straightforward approach to site evaluation, which has been in use since the 1960s, is to benchmark the performance of different locations on the basis of shared characteristics (Applebaum 1966). These models are therefore logically described as 'analogues'. An elementary approach along these lines, which has been adopted by at least one major high street retailer in the UK, is to simply create a matrix of locations in which the size of the retail centre is shown on one axis, and the size of the retail outlet is shown on the other. Groupings of the sites in this matrix now indicate locations at which similar levels of performance can be achieved. This method can also be a convenient way of determining the store layout and product offer at a location.

The obvious problem with the analogue modelling approach is that it may be difficult to capture the variation in outlet performance with just a small number of quite straightforward variables. For example, in working with a UK regional building society, centres were scaled on the basis of the retail strength at each location. Three East Anglian towns in the UK (for the record, Swaffham, Thetford and Dereham) were all identified as being equally strong. In one of these centres, a branch with 71 sq m of space generated £3 million of income, while a second with 115 sq m generated only £2.5 million. The third of the branches was of intermediate size – 96 sq m – but accrued more than £4 million in turnover. All of this despite the fact that across the whole network there was a strong general relationship between branch size and income generation (r-squared = 0.29).

To make this approach a little more convenient, further axes can be added to the matrix; for example, to account for varying levels of affluence in a town, or the quality and prominence of the site. The method can also be extended by weighting the variables relative to one another: according to Mendes and Themido (2004: 5) 'a large Portuguese retail group uses this method'. The further the models are extended in this direction, the more this approach starts to blend in with multivariate regression, which may be considered as the standard technique for estimating the relationship between a dependent variable and any number of potential correlates or predictors. Thus if we are interested in trying to estimate the turnover of a retail outlet (the dependent variable) on the basis of all sorts of factors such as those discussed in the previous section; that is, store attractiveness, footfall, car parking, etc., then this looks like a natural technique.

The classic weakness of regression modelling for store location analysis is that it is a multidimensional analysis based on store-level characteristics (see again Chapter 5). Therefore, it is likely to be much better at allowing for provision factors (in the terminology of the previous section) than either demand or technology, but the relative advantage of using a more sophisticated approach like SIM may be less obvious in the complex markets that we are considering here. Furthermore, there may be ways in which the other factors can be proxied. This is increasingly the case now that retailers are able to access relatively high-quality data about their customers, for example from either loyalty card data or third-party market research and lifestyle surveys.

In summary, there are a number of alternative approaches which suggest themselves as candidates for modelling complex markets, although the case remains to be proven that any of these are fully satisfactory. In the remainder of this chapter, we consider two methods that have been developed further in our own applied work. One group of methods is based on the idea of *rating* stores, but first we will consider approaches based on the *segmentation* of outlets.

9.3 Segmentation

As we saw in Chapter 4, the idea of segmentation has been applied enthusiastically to the problem of customer differentiation. In this context, some form of cluster analysis (usually k-means) is used to distinguish on a wide basis of socio-economic and demographic factors. Other techniques such as genetic algorithms are also increasingly popular in this process. The essential idea is that different geodemographic segments can be associated with different purchasing patterns and behaviours.

The concept of segmentation translates quite naturally to the issue of store assessment. Here the problem is to group the stores on the basis of multiple characteristics in order to capture performance variations and distinct market opportunities within each group. Some of the major factors which might be considered here include location factors (e.g. footfall, quality of parking, pedestrianisation, accessibility), competition factors (the quality, size and range of competing outlets, perhaps also including positive externalities such as retail adjacencies), demographic factors (spending in the catchment area, factors affecting brand preference such as age and affluence), and site factors (floor space, frontage, layout and so on).

As part of a network representation study for the UK Wrekin Building Society (a small regional financial service organisation) a segmentation system was developed by consultants from the University of Leeds. The objectives of this process were twofold: first, as a basis for configuration management in the existing network, in order to identify the ideal format for each outlet (as it was hypothesised that different branch types would be most appropriate and effective according to the local situation); and second, as a basis to evaluate opportunities for network expansion by allowing comparisons of new centres to those within the existing network. In this case, the characteristics of each outlet were grouped into three component indicators to measure the strength of the network, the strength of the centre, and the quality of the catchment area (i.e. the strength of demand). In each of these cases, multiple component factors were taken into consideration. For example, to assess the strength of demand, population density, household composition, expenditure by age, gender and social group, and the geodemographic classification of the catchment area were all taken into consideration.

The segmentation was created by splitting the network and centre components into three (high-medium-low), and the demand component into two (high-low) leading to the construction of 18 segments. From this analysis it was clear that network strength is a dominant and consistent driver of sales performance (for equivalent centre-demand combinations, performance is always better in clusters of high network strength – that is, cluster 1 exceeds cluster 7 exceeds cluster 13; cluster 2 exceeds cluster 8 exceeds cluster 14; and so on); and also that small centres with modest demand often yielded good opportunities without necessarily requiring heavy investment in the branch infrastructure. From this analysis, the inference was that the way to expand the network most effectively was not down the retail hierarchy (by expanding into the largest available centres in the surrounding areas), but by extending the network across a full range of centre types to achieve network saturation across a specific geographical region (e.g. by expanding from UK county Staffordshire into Derbyshire, Cheshire or Shropshire *but only one of these*). From the second phase of the analysis the type of representation needed within a target centre was identified according to the 'decision tree' as illustrated in Figure 9.3.

From this example, we can characterise the approach as the construction of a 'tree of alternatives' and in many cases it is possible to use automated statistical and modelling techniques to assist in this process, of which the best known is the chi-squared automatic interaction detector (or CHAID) which is readily available as an add-on to SPSS and through other statistical packages. The formal basis of this approach is to take a dependent variable and a series of independent variables, and by splitting each of the dependent variables to find which is the best predictor of variations in the independent

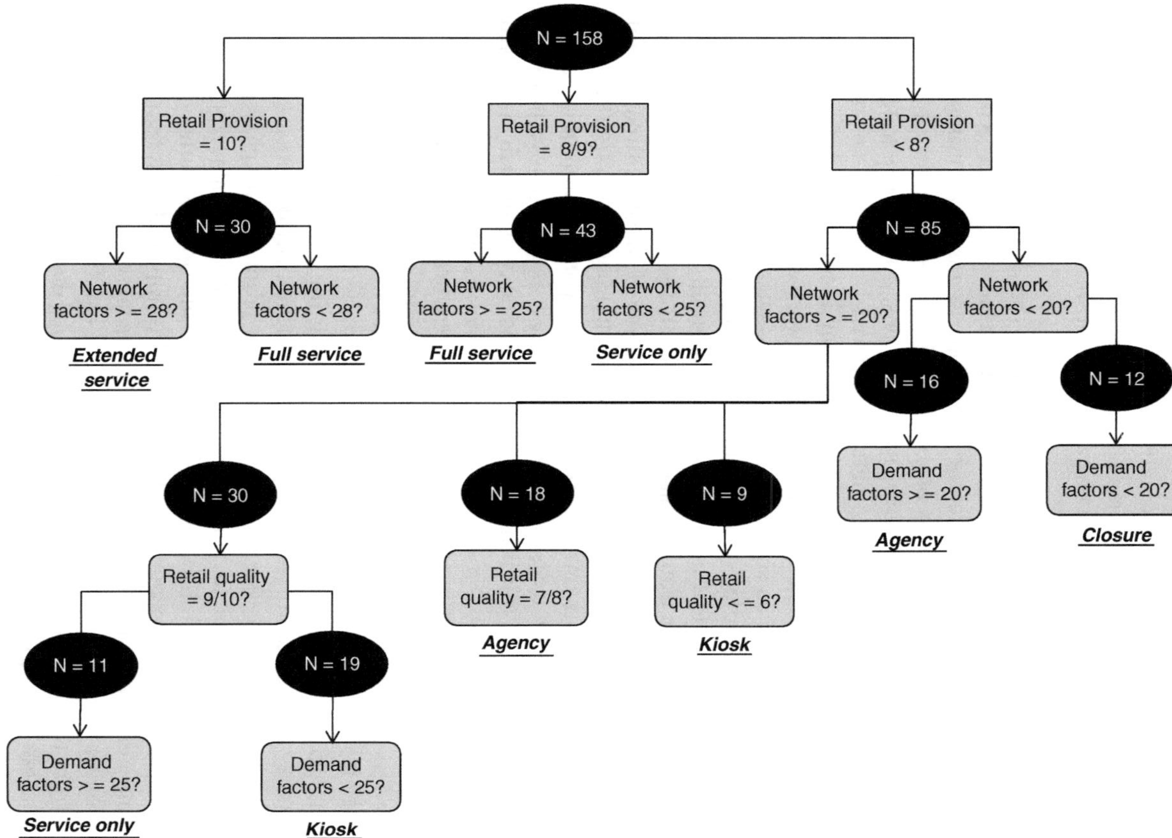

Figure 9.3 Decision tree for the Wrekin Building Society.
Source: Authors

variable. This split in the first independent variable (retail provision in the example of Figure 9.3) creates the first pair of branches in the tree. Each branch of the tree is then extended outwards in the same way, until further significant differences in the independent variable can no longer be identified. In the larger centres, full service branches are generally required, but this offer can be restricted where the network coverage is weak. In small centres, where the networks are weak, then closures can be justified, except in certain instances where pockets of concentrated demand may justify limited service through an agency.

In practice, one drawback of this approach is that it assumes a single independent variable and often it will be more desirable to consider performance as a much wider basket of indicators. Another issue is that automated techniques are difficult to apply to problems with a small number of observations – again as in our previous example with only 158 branches. Third, and perhaps most importantly, automatically derived trees are often messy and confusing. For example, different branches may incorporate different numbers and types of variables. In practice, therefore, we suggest that it is better to present the clusters as simple, exhaustive classifications of a network (e.g. a branch is either large or small, in an area of high or low population) while using whatever statistical approaches are helpful as an aid to the classification process (i.e. in the example above, we present the network in 18 clusters, produced from the combination of three indicators of network strength, three indicators of centre size, and two indicators of market demand – the derivation of the indicators themselves is informed by a combination of multivariate regression, decision trees, k-means cluster analysis and other techniques as appropriate, but all of this is hidden from the end-user in an easily intelligible final product).

Our Wrekin Case Study has illustrated that segmentation is a useful basis for estimating variations in sales performance in a retail network, and this can be used as a way to infer the likely sales volumes of new branches and outlets. It also provides a means for benchmarking the extent to which existing sites are realising their potential, hence as a basis for setting sales targets and quotas, as well as identifying category leaders (high performance branches) from which lessons in effective management or promotion might be drawn. Studying the variation in performance of different branch layouts across different segments can also serve as a basis for the optimisation of formats to the local context. In our own experience, this approach has been applied extremely effectively to problems involving financial services branch networks and convenience store networks, both of which can be classed as hard problems from the perspective of SIM, as we explained earlier in this section.

Another example of network reinvention is work undertaken by the authors for the UK Post Office. Traditionally the Post Office served as a major provider of welfare benefits paid over the counter. These included services such as pensions, unemployment benefits, child benefits and so on. However, over the last 20 years these services have largely been automated, so that payments go directly in to recipient's banks account and the Post Office has had to develop a new business model becoming more like a twenty-first-century retailer than a benefits outlet. Under the terms of its agreement with the UK government it has a social obligation to make outlets available to 95 per cent of the UK population within a ten-minute drive-time. This led to the development of one of the UK's largest network of branches. At the time the analysis was undertaken, the Post Office had around 14,500 outlets. As a benchmark that is more than the combined network of all bank branches, car dealers and large superstores in the UK. The network is divided into two types of branches. The first is directly owned and managed outlets (often termed 'Crown' branches) that tend to be located in cities and larger towns. These total around 1,200 of the 14,500 outlets: the staff are Post Office employees and they focus solely on Post Office products. The remaining, and bulk of the outlets, are essentially agents (franchises) of the Post Office which combine providing counter services for Post Office products with other retail services. Some of these are independently owned businesses, some are part of symbol groups (e.g. Spar, Londis) and some are part of national retail groups, such as WHSmith. The network is therefore incredibly diverse. In addition, 61 per cent of the total Post Office revenue is generated by just 10 per cent of the branch network and some of the rural branches have very low sales. Not surprisingly therefore, the viability of the network has often been called into question and there has been a wave of branch closures over the last decade.

The project we undertook was a 'clean slate' review of what the Post Office network could look like in the twenty-first century (financially viable but continuing to fulfil its social obligations). Our initial analysis quickly concluded that the existing network was commercially unviable. While it was identified that 19 per cent of the UK's population lived in rural areas, 50 per cent of Post Offices were located in centres classified as 'rural'. To develop a recommendation on what a future network might look like we developed an eight stage framework for the analysis:

1 Define CMAs.
2 Outlet scorecard by product created.
3 Segmentation of existing network.
4 Network blueprint built.
5 Format optimisation undertaken.
6 Location and format for a viable network presented.
7 Optimal channel mix by CMA defined.
8 Results reported.

The first stage thus involved the development of Customer Marketing Areas (CMAs) that have already been discussed in Chapter 8. The CMA framework allowed us to break down the UK in to manageable planning areas. Figure 9.4 shows the CMAs defined for south-west England and south

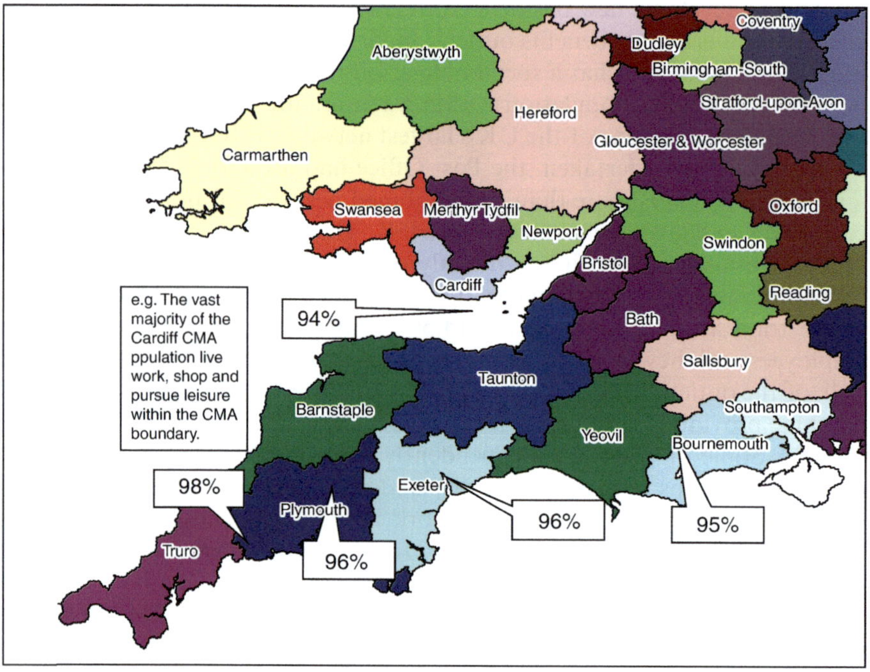

Figure 9.4 CMAs in south-west England and Wales.
Source: Authors

Wales, demonstrating very high levels of containment (i.e. where people live, work and shop).

The next step was to examine the performance of existing branches in the network. To achieve this we had to undertake a segmentation of the network into branches of a similar type. Given that branches vary from large city centre outlets to small rural branches it is important to compare like with like. We would not expect a small branch to generate the same levels of sales as a large one, but the key is to find, for each branch, whether they are under or over-performing given their location. From there we developed a network blueprint for each CMA based on two criteria: profitability and customer coverage. We then examined what products would be most appropriate to be sold in each branch. This came together in the final stage with recommendations for a set of locations and formats for a viable network.

To put together this framework we needed to access a wide variety of data sets. These included:

- small-area population counts (2001 Census)
- market research data on spending patterns

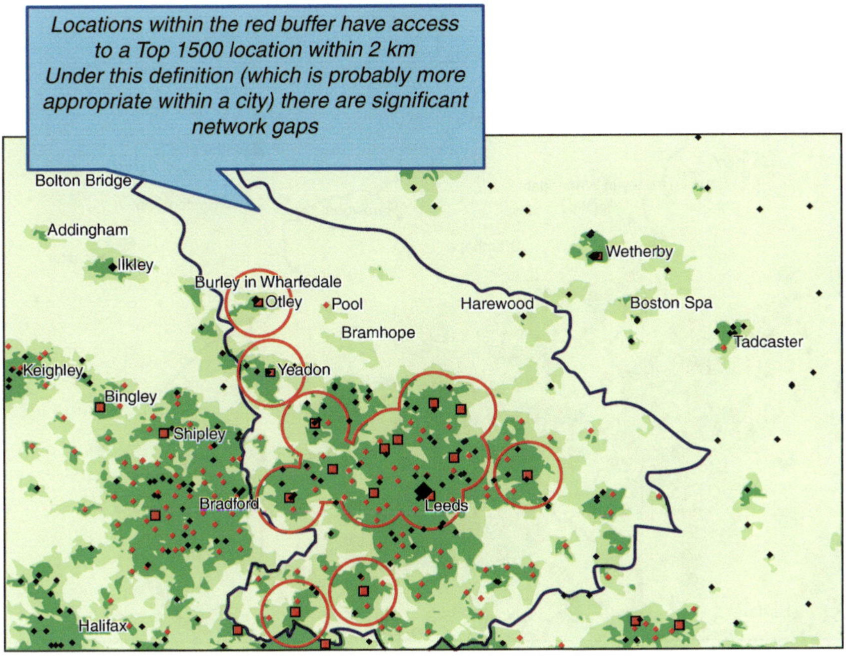

Figure 9.5 Network provision for the Post Office in Leeds CMA.
Source: Authors

- competitor retail locations
- Post Office branch sales data
- Post Office channel transactions database
- industry channel research
- Census journey to work data
- employment data
- ONS population projections.

The results of our analysis suggested that a two-tier network would provide an 'optimal' solution to producing a viable network that would also meet the Post Office's social obligations in terms of consumer access. This network would comprise 1,500 branches that would be located in main urban centres and another 1,500 branches that would provide the 'infill' to meet the accessibility criterion. Even though a reduction from 14,500 branches to 3,000 seems quite dramatic this would still leave the Post Office as an organisation with a large retail network in the UK.

To illustrate these results for two different types of CMA we use the examples of Leeds and Exeter. Leeds is largely a metropolitan CMA. Figure 9.5 shows the location of the top 1,500 branches in red and the 2 km buffer zones

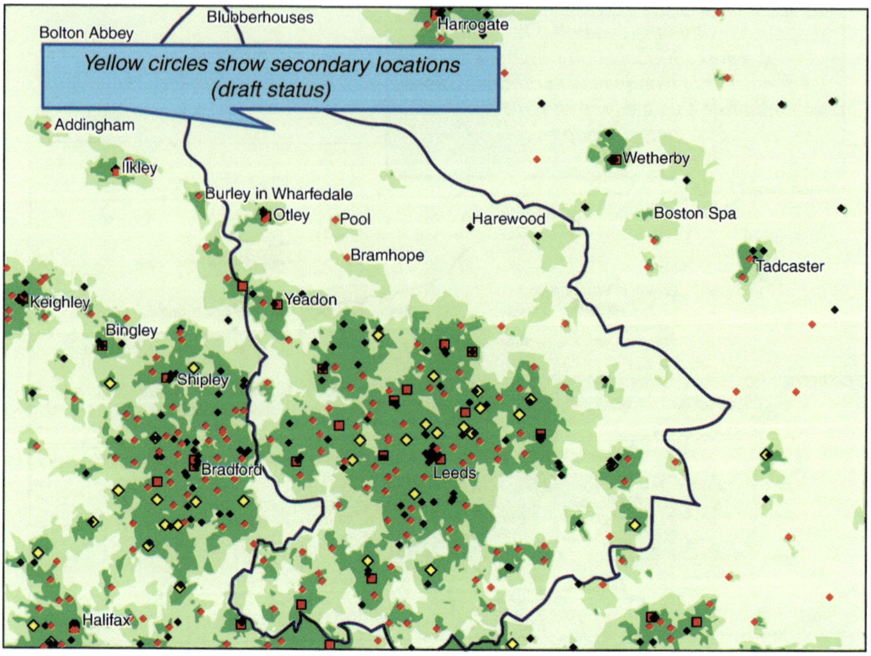

Figure 9.6 Filling in the gaps in network provision for the Post Office in Leeds CMA.
Source: Authors

around this network in Leeds. As can be seen there are still quite significant gaps in network coverage so on Figure 9.6 we identify a secondary network of branches, shown in yellow, that helps fill these gaps.

The Exeter CMA is a mixed urban/rural CMA with a small number of large/medium sized towns but quite a large rural hinterland. Figure 9.7 shows our recommendations for the primary network along with associated 5 km buffers. There are still considerable gaps in coverage so we had to develop a secondary network of branches to make good this shortfall. These are shown, alongside the primary network in Figure 9.8. This secondary network may well have a specialist product focus, limited opening hours or other restricted service offerings.

9.4 Site rating models

If we wish to benchmark the relative performance of a retail outlet in the centre of a city such as Leeds in the UK, then it makes much more sense to compare against outlets in centres of other major UK centres such as Birmingham or Manchester, than against smaller market towns such as Otley or Ilkley (small market towns close to Leeds). In essence, this is what

Figure 9.7 Network provision for the Post Office in Exeter CMA.
Source: Authors

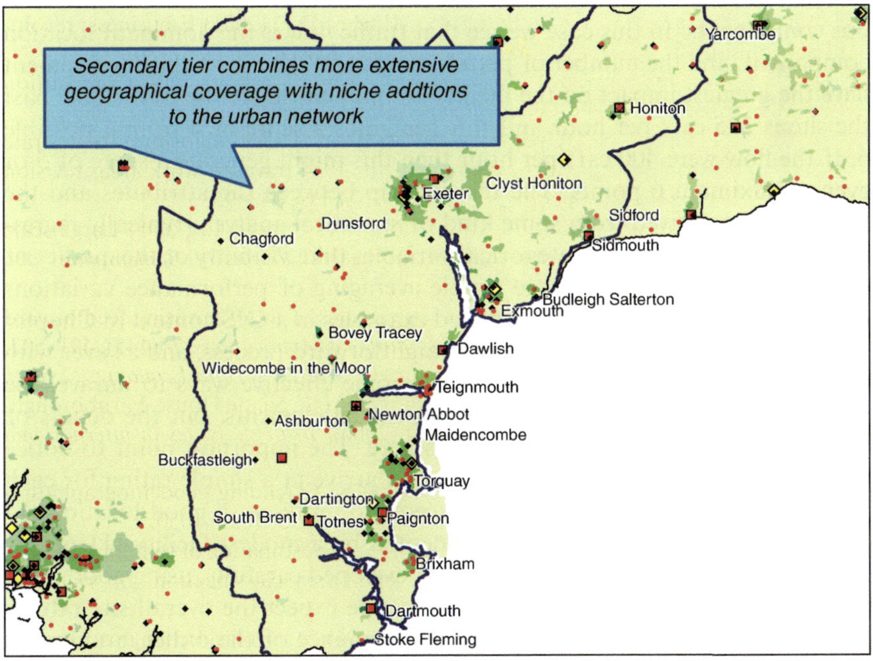

Figure 9.8 Filling in the gaps in network provision for the Post Office in Exeter CMA.
Source: Authors

we achieve through the segmentation approach. But what happens if the Leeds store is actually twice the size of its Manchester counterpart, and is in a prime retail development next to Marks and Spencer, rather than in a slightly faded 1960s unit on a peripheral street with many boarded up properties to let? In order to make this distinction, we might think it wise to qualify our estimate of outlet type with a more precise attempt to quantify the advantages of each outlet. This is what we are seeking to achieve with the rating model, and from this discussion the complementarity between the two approaches can be clearly seen.

Let us consider immediately the example shown in Figure 9.9 which shows outputs from a site rating tool used in Yorkshire, UK (the tool was originally developed some years ago for the petrol market in Ireland). The rating is built up from two bundles of attributes: on the left-hand side, those relating to the location of the outlet, and on the right-hand side those relating to the facility itself. The location attributes are fundamental characteristics of the site which are outside the control of the retailer. Sometimes these features are referred to in the petrol industry as 'dirt strength'. In choosing a site on the basis of location attributes, the developer wishes to understand the relative importance of obtaining a high rather than a low traffic flow. Facility attributes are under the direct control of the developer of the site. Here the decision to be made is whether the site requires six fuelling positions or eight, and thus whether any increased benefit outweighs the cost of the investment.

The scores in the rating model reflect the relative importance of the various components. In this case we see that traffic flow is the dominant location component, and the number of petrol pumps and the age of the development have the greatest impact on the facility component. Here the traffic flow past the site is 225 cars per hour and this generates a score of 4 from a possible 6. If the flow were 400 cars per hour then this might generate a score of 5 or even a maximum 6 points. The relationship between the attributes and the scores can be derived from some kind of statistical analysis, typically regression, or in the case of the categorical variables (like visibility or the quality of the facility) then through some simple averaging of performance variations between the categories (see also related examples in a GIS context in Chapter 3). Of course this is not an entirely straightforward process, and as ever with multivariate analysis the trick is to find some effective ways to unravel the interdependencies between the different model elements, but the details of this process do not need to concern us here. The important point to notice is that at the end of the procedure, we can arrive at a simple rating for each site. In this case the score is 25, and seems to represent a good location (14 points from a maximum 20) complemented by a modest facility (11 out of 20). Assuming that our model has been developed carefully, using good data, and for a reasonably well-behaved market, we expect the overall site ratings to be highly correlated with the sales performance of the outlet, and in this instance we see that this is clearly the case, as indicated by the 'Performance' statistic which appears under the rating. Here the performance is 95 (index to 100) from estimated sales of 2,100 kiloliters per year, from which we can infer that the usual sales for a site with a rating of 25 would be little higher – in the order of 2,200 kilolitres.

Thus looking at Figure 9.9, what we can see here is an attempt to draw comparisons between sites based on a number of different attributes which all have relative scores, and in this sense the method does not seem very different to the creation of an analogue model with multiple inputs and in which each of these inputs is carefully weighted. Therefore this model may not be so different to the kind of device that is in use by the major Portuguese retailer described by Mendes and Themido (2004). Equally, however, there are some similarities to multiple regression in which the net effect is to compare the impact of different variables, albeit that the scorecard is able to capture categorical data and to represent non-linear relationships much more easily. The fundamental point, perhaps, is that there are overlaps and linkages between many of the techniques discussed in this chapter, all of which are trying to use best available data to address similar problems and questions.

While we have referred to this procedure consistently thus far as a *rating* system, the term *scorecard* has also been commonly applied to models of the type shown in Figure 9.9, and clearly this is a natural description. So it is worth noting that an alternative provenance for this entire approach

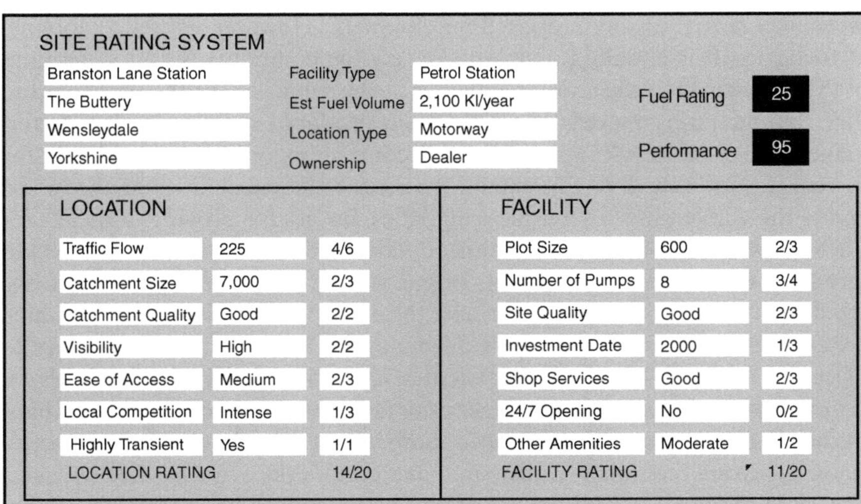

Figure 9.9 Site rating tool for a petrol station.
Source: Authors

is from the many scorecards that have been highly popular in business and industry for two decades or more, where for example the *balanced scorecard* became – for a while at least – something of a management ideal. The idea can also be related to the *league tables* which are so popular in government and public services today, for the comparison of hospitals, police forces, schools, and of course universities and their constituent departments. For example, typical University Guides will rank academic subjects according to their (supposed) attractiveness to undergraduate students (e.g. Guardian undated). The similarity in the procedures is obvious – a range of key indicator scores (employability, staff-student ratios and so on) are captured, weighted, combined and evaluated, and higher overall scores are associated with better performance. Hence the rating model connects across not just to analogue models and other retail assessment tools, but also chimes directly with a number of methods in common use in both public services and the business world. Credibility and ease of understanding flows naturally from such considerations.

Returning to Figure 9.9, it is also worth discussing the purpose of the rating tool in more detail. In this case, we can see that the object is to gauge the likely fuel sales of the (imaginary) Branston Lane site. Clearly an effective scorecard model has significant potential as a performance evaluation tool – where are the outlets in which sales are significantly below par, and what can be done about this? And which outlets are the star performers, and what can be learned from their management or promotional strategies? Second, and again in common with many of the other models discussed in this chapter and elsewhere in the book, the scorecard can be a powerful appraisal tool to determine the revenue potential of any proposed development. However, as well as looking at fuel sales, it is equally possible to review the components and weightings in the scorecard to reflect other activities at the same site. In the petrol sector, for example, rating models have also been produced to estimate the performance of a quick service restaurant, forecourt shop, or car wash units on the forecourt. Not only does this extend the assessment capabilities of the model across the whole range of business activities, but it also provides a means for the best use of retail space – if I want to redevelop the site, would it be best to introduce quick service restaurants or car wash, or to refresh and extend the retail facilities, or maybe just to increase the number of fuelling positions which are available. The solution to these dilemmas will typically vary from place to place – in suburban locations, the potential for both pedestrianised and vehicular patronage of the shop may be considerable, while on motorways perhaps the need for quick serve restaurants is more urgent. The development of petrol stations beyond their historically core function of selling fuel has been a major feature in recent years and the use of site assessment tools has been crucial in this process – for more discussion and examples, see Birkin *et al.* (2003).

In this section, we have spent some time on an example of the rating model as it has been applied to petrol stations for a variety of purposes. The downstream retail petroleum sector has many of the attributes discussed earlier in this chapter which make it difficult to develop more sophisticated sales models, especially non-residential convenience-based trips which often form part of much more complicated social and economic interactions. Similar models have also been successfully applied, for the same sorts of reasons, to the location of electronic cash machines (ATMs). In many of these contexts, it might be noted that poverty of data is a strong driver towards the adoption of simpler models, and in consequence it might also be argued that as data resources become increasingly rich then the value of these more straightforward approaches could be reduced. Against this, however, we should set the argument that rating systems are easy to develop and readily comprehended not just by analysts but by users and decision makers across a business, and these features in particular may justify the continued popularity of the technique.

9.5 Investment appraisal models

The last analysis approach that we will consider in this chapter is somewhat different to the others in addressing directly the challenge of translating the spatial and demographic impacts of network reinvention from a financial perspective. The key observation is a straightforward one, but nevertheless bears emphasis – the spatial and demographic impact of network revisions has a direct effect on the bottom-line of the business. A major British DIY retailer with a network of over 100 stores was unable to develop a convincing methodology for accurate sales forecasting. This led to a lack of confidence in site assessment capabilities, from this a failure to exploit new opportunities, and ultimately reduced expansion and limited market competitiveness. From a relatively modest investment in its network modelling capability, the retailer was able to reverse this process and hence achieve much more substantial benefits than simply a marginal increase in its sales forecasting capability (see also Birkin *et al.* 1996, for more details on this example).

The means by which a retailer might assess the financial value of a retail development are illustrated in Figure 9.10. Here we are considering a supermarket retailer that has the opportunity to expand its operations through the redevelopment of an existing site. This redevelopment process yields increased profits from a combination of food and non-food sales, but notice immediately that one of the problems faced by all retailers in this situation is that it requires money up front to introduce the possibility of extra returns sometime in the future. From the financier's viewpoint, cash in the bank now is worth more than the promise of cash in the future – in part because of uncertainties, and in part because of the opportunity cost of other potential investment strategies. The reduced value of future profits is recognised by adjusting them back to a net present value, and in this case this is achieved by reducing the value of projected profits by 15 per cent every year.

NPV Scenario Summary Report			
Annual Turnover	£	48,736,643	
		NPV at 15%	Average margin
Food	£	2,426,746	6.03%
Non–Food	£	1,072,545	9.15%
Other	£	–	0.00%
Total earnings	£	3,499,291	7.18%
Retail overheads	–£	547,499	–1.01%
Supply and distribution	–£	683,783	–1.44%
Property costs	–£	288,434	–0.65%
Third party fees	–£	373,624	–0.75%
Credit card costs	–£	282,552	–0.77%
Total costs	–£	2,175,892	–4.70%
Net income		1,323,399	2.48%
Net income	£	1,323,399	
Investment	–£	513,331	
Tax	–£	272,952	
Working capital	–£	75,598	
	£	461,518	
Net cash flow	£	461,518	
DCFR		15.2%	

Figure 9.10 Net present values for a retail investment.
Source: Authors

In order to generate these extra earnings, certain direct costs are involved (i.e. retail overheads and distribution costs, and also a proportion of retail sales that are attributed to credit card suppliers) as well as the charges associated with acquisition of property for the redevelopment, third-party charges to estate agents, solicitors and so forth. The deduction of these costs from total earnings gives the net income from the project. Unfortunately, the accountants have still not finished as there remains a need to allow for the tax on profits, the working capital requirements of the redevelopment process and the investment charges accrued in raising funds for the project. At the end of all this, the overall returns from the project can be expressed as a Net Cash Flow which measures the financial surplus from the project and which may also be represented in percentage terms as a Discounted Cash Flow Rate (DCFR).

These details have been sparsely presented on purpose as they are not germane to the main focus of this book and the interested reader can find out more about the underlying financial procedures from a more specialist source (e.g. Pike and Neale 2003). The crucial point to note is that the DCFR or rate of return on the project is ultimately determined by the earning capability of the redevelopment, and it is this earning potential, which we have argued above, can only be determined with any degree of confidence through the appropriate use of spatial analysis techniques. The accountants and investors in any retail business will typically expect a rate of return which is appropriate to the economic climate of the time – let us say in the current case that this rate of return is 15 per cent, and so it seems that this project just passes the test of viability.

These ideas form the basis for the Investment Appraisal Model which is shown in the next illustration (see Figure 9.11). Here we see the model has been configured to link to the previous example – thus there is a project that creates earnings of £461,500 with a rate of return at 15.2 per cent (see boxes at top centre left). From the box on the right, we can see that lower discount rates make this project seem even more favourable, but any further increases beyond 15 per cent will reverse its viability. Of more interest, however, are the boxes on the left which explore the sensitivity of the project to variations in the input assumptions. Here we are considering the possibility that retail sales have been overestimated on either the food or non-food sides, that the investment costs are greater than expected, or that the profit margins on trading are too optimistic. Given the marginal nature of this project, it is not surprising to see that a more conservative view on any of these inputs can render the project infeasible.

To conclude, model evaluations of a retail network or its constituent sites are intimately linked to the financial appraisal of network reinvention projects.

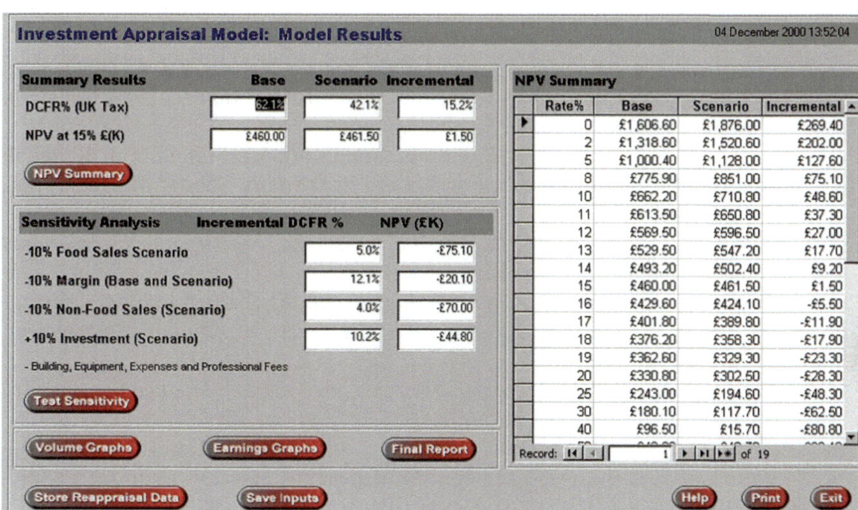

Figure 9.11 Outputs from an investment appraisal model.
Source: Authors

Many retail businesses employ legions of accountants and financial controllers to monitor their financial affairs. Surely they would be well advised to pay equal attention to the creation of models which can provide best available estimates of the impacts of retail network change.

9.6 Summary and conclusions

This chapter has provided a detailed overview of the underlying causes and effects of many years of turbulence in retail markets. We have attempted to articulate with some care the proposition that the overall outcome in many cases is a set of customer behaviour patterns that are difficult to model using the technologies that have been favoured elsewhere in this volume. In such cases a range of other methods have been reviewed and assessed. The contribution of three specific techniques have been given specific attention, and the value of these approaches has been explored in relation to strategic planning (network segmentation), store formats and stocking (site rating) and financial evaluation (investment appraisal models).

10 E-retailing

10.1 Background and introduction

The terms 'e-commerce' and 'e-business' have been in use since the mid-1990s, for example, with the first appearance of the Journal of Electronic Commerce (Zwass 1996). Related terms such as e-service(s), e-retailing, (and even e-trust and e-satisfaction), online shopping, bricks and clicks, can all be found abundantly in the academic literature. While the distinction between these terms is sometimes subtle, the primary focus of the current chapter will be on e-retailing, for which we will follow the working definition of Mokhtarian (2004) of 'searching and/or purchasing consumer goods and services via the internet'. E-commerce and e-business will both be regarded as somewhat broader in their scope: for example, extending to include supply chain logistics and business process management; online shopping will be taken as equivalent unless otherwise stated and other variations on the concept will be defined explicitly as necessary. In particular, towards the end of the chapter, we will consider even more novel phenomena such as m-commerce and s-commerce, paving the way for more detailed discussion of these in Chapter 11.

Going back to the early days, commentators have questioned the long-term impact of the internet on business processes (Coltman *et al.* 2001). Whether these changes might most appropriately be classified as 'revolution', 'evolution' or 'hype' is a topic that might still be debated at considerable length, but disregarding semantics it is clear that e-retailing has had a profound impact on the sale and purchase of consumer goods in the second decade of the twenty-first century. In purely statistical terms, e-retail activity has grown rapidly to substantial levels. A study in 2015 by the UK's Centre for Retail Research estimated that retail businesses generated an estimated £44.9 billion from e-commerce sales in 2014, with a forecast for £52 billion by the end of 2015. Thus, in 2014, market share for e-commerce sales in the UK was reported at being around 15 per cent of total retail sales. In the USA, the equivalent figures were reported at $306 billion in the year to October 2014, producing a channel share around 7 per cent and with an annual growth rate of more than 15 per cent (US Census Bureau 2014; Centre for Retail Research 2015). China is now recognised as the 'biggest digital market in the world' (China Internet

Watch 2014) with total retail sales equivalent to US$390 billion, a 9.6 per cent channel share growing at three times the market average.

A somewhat controversial issue in the growth of e-retailing has been the impact or importance of geography. We will seek to demonstrate below that the famous declaration of the 'death of distance' (Cairncross 1997) was not merely overstated, but simply inaccurate. Relative location in space remains vitally important in the analysis, evaluation and planning of retail location networks even for e-sales and continues to influence customers, products, channels and distribution networks quite deeply.

The remainder of this chapter is structured as follows. In Section 10.2, we will consider a range of issues relating mainly to the supply-side of the retail process. Demographics and retail demand will be the focus of Section 10.3. The attention will shift to the interactions between consumers and retailers in Section 10.4, in which we will add more substance to the argument that understanding spatial patterns remains as important as ever. The final section will consider the influence of the latest developments in technology and thus also provide some pointers to future directions of change.

10.2 The supply-side

The extent to which established retail businesses are under threat from online adversaries has been a topic of some debate in the literature. The British supermarket grocer Tesco is one organisation that has been able to use electronic retailing as a means of strengthening its hold on the market. According to Wilson-Jeanselme and Reynolds (2006), by 2005 Tesco was operating the largest online grocery retail business in the world, with a presence in Ireland, South Korea and the USA as well as its native UK. Between 2013 and 2014, Tesco's online UK grocery sales rose 12.8 per cent to £3.2 billion. While the online service is packaged from many of Tesco's large supermarkets it now has five 'dark stores', where employees assemble customers' online orders to serve its web business (a sixth was planned to open in 2014–15; see *Guardian* 2013). At home, the business enjoys a 37 per cent share of the online grocery market (2014). In contrast to the customer base of its supermarket outlets, 80 per cent of Tesco online customers in 2006 were reported to be aged between 25 and 49, and well over half (55 per cent) were men (Wilson-Jeanselme and Reynolds 2006: see the further discussion in Section 10.2).

Tesco's online business has continued to prosper in recent years and now sells more than one billion items online each year. Its core product offer is complemented by specialist sites for products such as clothing and entertainment. Hence, overall, Tesco has been able to make a success of its online business by the integration of e-retail activities alongside mainstream business activities, even to the extent that the stores are shared at the heart of the distribution network. Such models of electronic business integration have not always been viewed so positively, both in theory and in practice. For example, the financial services giant Prudential moved decisively at an early stage to establish an independent e-business subsidiary under the Egg brand. According to Enders and Jelassi (2000) the separation of electronic operations creates a unique opportunity for enhanced corporate performance through 'the ability to create an entrepreneurial atmosphere in (the) online business division'. In other words, the benefits of internal competition for customers, combined with a certain nimble-footedness and a fresh approach are seen as the key determinants of strategy. Gaining trust is paramount also to that successful strategy (Mortimer *et al.* 2016; Pappas 2016).

A more balanced approach to the problem of integration versus separation is presented by Nicholls and Watson (2005), in which the most appropriate strategic response is determined by the positioning of a company's activity in relation to three 'situational antecedents'. These antecedents are the strategic objectives, business characteristics and resources and competencies of the organisation. Strategic objectives could include sales growth or cutting costs; extending the geographical range of the brand, product differentiation or extended penetration of new markets. For example, if the main objective is to cut costs then business integration is the obvious way forward, but if the focus is to maximise the recruitment of new customers then separation may be a more promising way forward. Relevant business characteristics could include the size and location of the firm, the product sector(s) in which it operates, the strength and quality of its brand, and the nature of its competitive environment. For example, a regional chain might favour the separate development of an online presence as a means for wider expansion, while an established national player might prefer to focus on consolidation of its brand through a more integrated approach. The resources and competencies of an organisation could comprise its corporate technology infrastructure including IT hardware and databases, distribution and logistics capacity, and less tangible qualities such as management skill and flexible learning abilities. Major retail organisations (like Tesco or one of its major competitors, Asda, Sainsbury's or Morrisons) all have sufficient store capacity to pick and deliver within an integrated business, while a smaller competitor such as Waitrose has preferred to focus its attack on the e-retail marketplace through joint operations with a specialist partner, Ocado.

In general, however, Nicholls and Watson (2005) are critical of the quality of the strategic response of British retailers to the opportunities afforded by e-commerce. From a survey of the top 500 retailers, they concluded that most had been reactive rather than proactive, typically offering limited online product ranges and tapping into the same customer niches. It is more by good luck than good judgement perhaps that these organisations have been able to survive and even prosper within a changing environment, since as argued by Gulati and Garino (2000) regardless of specific market and business conditions 'the benefits of integration are almost always too great to abandon entirely'. Or to put it another way, established retailers have such a head start that they will almost always be able to protect their position against new entrants. On a similar note, Moore and Ruddle (2000) 'believe the end point will be a fully integrated business'. Wilcocks and Sauer (2000) recognise that in the early days some degree of internal separation may be beneficial as a web presence is created, perhaps powering an independent sales operation, but only setting the scene for later integration of processes and ultimately leveraging the benefits of new (virtualised) capabilities to maximise overall company value.

Above all, perhaps, Tesco's experience is manifest evidence that retail intermediaries are far from redundant in the emergent ecology of electronic retailing. Writing in the late 1990s, Stern (1999) argues that e-retailing has low barriers to entry, making new businesses relatively easy and inexpensive to start; that e-tailers can get products to markets more quickly and react to market trends faster than their monolithic physical counterparts; carry broader assortments without the constraints of the built environment; and can exploit the capability for virtual product reviews and price checks. The natural outcome of the e-business revolution might therefore be seen as one of disintermediation, in which the retailer no longer has a legitimate filtering role in facilitating the acquisition of products from their manufactured source.

Consider an idealised representation of the retail system, as in Figure 10.1. The essential function of the retail store network is to mediate the supply chain from manufacturers to customers, by co-ordinating a range of products in a single location (Figure 10.1(a)). In a world of ubiquitous information flows across the internet, customers can now go straight to the manufacturer for the acquisition of new purchases, simplifying the chain and cutting costs for both the supplier (to establish and maintain a retail network) and for the customer (in accessing the stores) – see Figure 10.1(b). One of the disadvantages of this process is that the function of retailers to provide an important comparison function between alternative manufacturers and their respective brands and products is lost. As electronic markets have become more sophisticated we have therefore witnessed the emergence of new online aggregators or cybermediaries – companies like Kelkoo, GoCompare, LastMinute, Expedia and dabs.com – which provide comparisons of price (and to some extent quality) across the range of available merchandise. This process can be characterised as one of reintermediation (Figure 10.1(c)).

In practice, the onset of disintermediation has varied considerably between different retail sectors. Jan Weltevreden and colleagues at the University of Utrecht have undertaken intensive analysis of the Dutch Multiscope retail panel to detect trends across more than 20 product categories. Overall, the conclusion is that '[i]nternet users often purchase directly from manufacturers or service providers, *bypassing retail intermediaries only for a limited number of products* (travel, software and computers)'. This tends to counter earlier claims that the internet would lead to a disintermediation of retailers (Weltevreden 2007: 205, our italics).

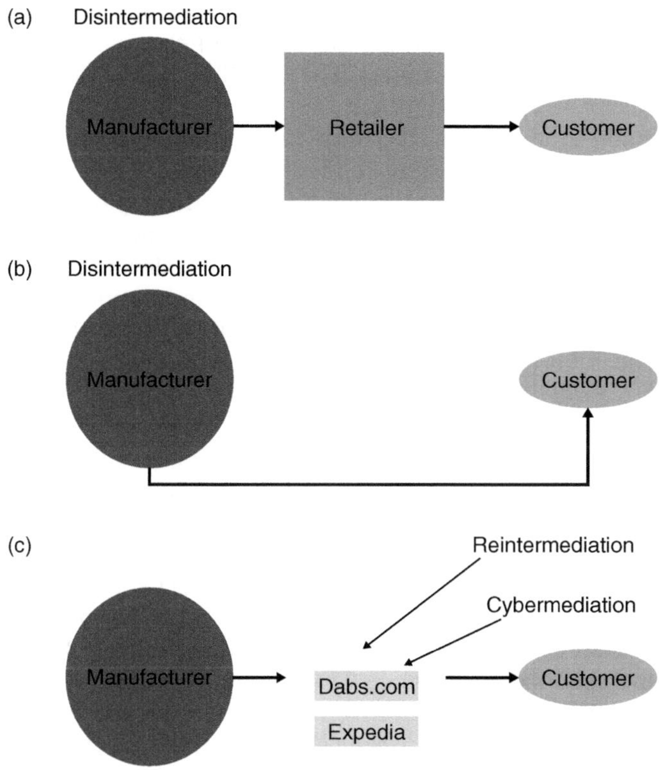

Figure 10.1 Dynamics of the retail process of disintermediation.
Source: Authors

A better understanding of these patterns can be gained if we consider some of the disadvantages of e-purchasing. Of course the internet denies customers the social pleasures of shopping (as noted by Stern 1999 and Reynolds 1997), which is especially important for comparison goods such as clothing, toys and gifts. More importantly, perhaps, the function of the retailer as an aggregator of products remains significant. Typical retail outlets feature many thousands of products and brands, and individual shopping baskets will feature tens or hundreds of individual items. The selection of individual products from its manufactured source would only be feasible where the purchases are relatively infrequent and of a high value (as in computer hardware, for example). Internet search is also most likely to be effective for products that are of an homogeneous quality (for example, books and CDs are of a uniform standard; fresh vegetables and cuts of meat are not). Finally, some products are by nature electronic, hence travel reservations or computer software can be readily exchanged through the fibre optics of the internet; cosmetics and items of jewellery cannot.

For these reasons, among others, grocery retailers have been more resistant to the onset of e-commerce than some others. However, one outcome of the internet is that it has made customers more aware of the convenience and time-efficiency of electronic transactions. The net effect is a progressive shift towards electronic distribution, but at a slower pace and under the control of established retailers rather than cybermediated usurpers as in some of the other sectors noted above. Some major trends at the category level in the UK retail sector can be drawn out from Acxiom's UK Research Opinion Poll (ROP) on expenditure patterns of more than a million consumers between 2007 and 2010 (G. Clarke *et al.* 2015). This data shows that 44 per cent of consumers now purchase books online with products such as wine around 25 per cent. Although groceries start from a much lower base than books and wine, growth has been considerably more rapid in recent years. According to Mintel (2013) online grocery sales at constant prices are expected to continue to grow explosively from £4.39 billion to £11.69 billion between 2010 and 2017. It seems certain that e-tail distribution will continue to be a major battleground between the major supermarket retailers for the foreseeable future.

10.3 The demand-side

In 2005, the market research organisation Acxiom asked over 600,000 consumers about their e-retail habits. At this stage only about 2 per cent of customers stated that they had shopped online. However, this average masked tremendous spatial diversity. Considering that UK postal areas are ranked from 1 to 120 according to the uptake of online shopping, there is a range of more than six times in e-uptake between the ten lowest ranking postal areas (average seven online customers per thousand population) and the ten highest ranking postal areas (average 44 online customers per thousand population). The lowest ranking areas are all to be found in the northern and western extremities of the UK – in Scotland (in particular the Shetlands, Highlands, Inverness and the Orkneys) and in Wales and the Welsh borders. In contrast, the seven highest ranked areas can all be found in Greater London with the next three all in close proximity – Reading, St Albans and Croydon.

In some ways, this pattern might be considered paradoxical. After all, it is customers in rural areas who have most difficulty in accessing high street retail outlets, and might therefore be considered as the group who have most to benefit from the provision of online alternatives. Farag *et al.* (2006) refer to this as the efficiency hypothesis, stating that 'consumers with a relatively low shop accessibility … have access (via the internet) to a larger variety of goods and services and can save both travel time and shopping time'. To account for the possibility of non-conformance to this pattern, they postulate an alternative innovation-diffusion hypothesis 'that new innovations follow a conventional pattern from large to small settlements' (Farag *et al.* 2006). This idea of the hierarchical diffusion of innovations can fairly be stated to be 'conventional' given its deep embedding within the geographical literature in relation to phenomena ranging from infrastructure (Taaffe *et al.* 1963) and culture (Morrill 1968) to diseases (Cliff 1981); see also Ferguson *et al.* (2005), or Epstein (2009) for a more recent twist on a similar theme.

In their own work, Farag *et al.* (2006) analysed data from 100,000 customers in the Dutch Multiscope panel, of whom 2,109 were active in e-retailing. Respondents were classified according to five levels of population density. They also found a higher propensity for consumers in strongly urbanised areas to search for and buy products online. However, these patterns were somewhat less starkly defined than in the UK Acxiom data described previously. Indeed, while the relationship for searching was statistically significant, the relationship for buying online was not, although to some extent this finding is affected by the small sample size in this investigation.

In thinking about the underlying mechanisms for both the efficiency and innovation-diffusion hypotheses, then the major factors to consider are demographics, access to technology and access to retail networks. In the remainder of this section we will discuss demographics and internet provision as potential drivers of the innovation-diffusion process. Access to retail networks and its influence via the efficiency hypothesis will be reserved for the next section on spatial interaction.

Demographic analysis of online shopping patterns in the UK Acxiom ROP data is continued in Figures 10.2, 10.3 and 10.4. This section draws on the recent analysis by G. Clarke *et al.* (2015). Figure 10.2 gives an age breakdown of electronic purchasing according to four response categories – customers who 'never' buy online and 'rarely' do so, versus customers who 'sometimes'

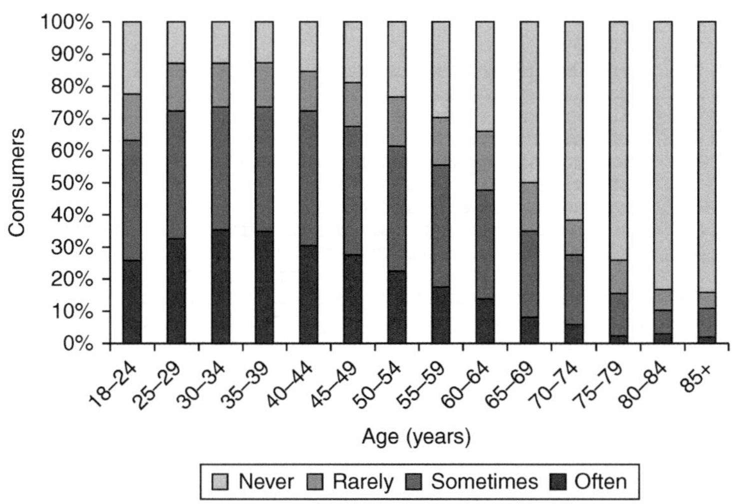

Figure 10.2 Demographics of e-retail (age).
Source: Clarke *et al.* (2015)

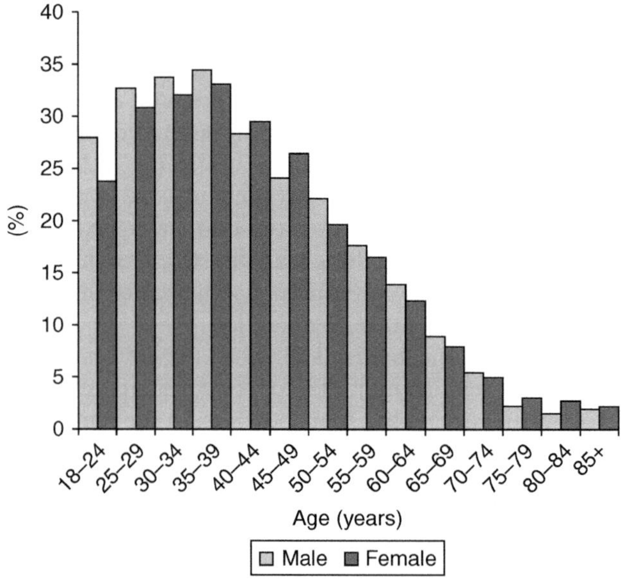

Figure 10.3 The demographics of e-retail (age and gender).
Source: Clarke *et al.* (2015)

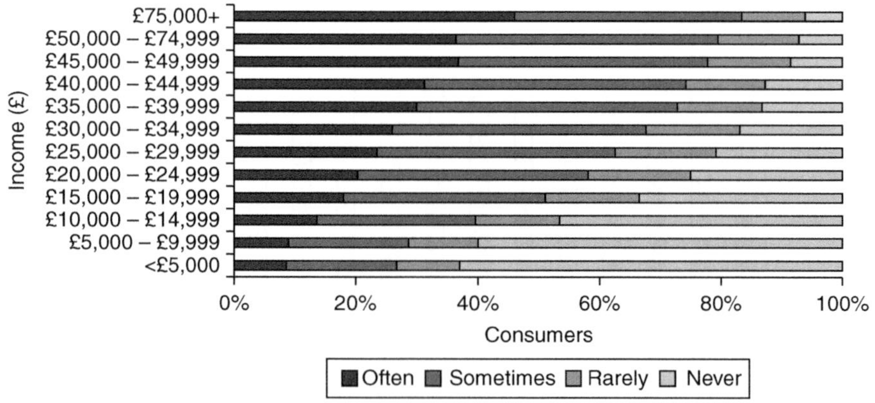

Figure 10.4 The socio-demographics of e-retail (income).
Source: Clarke *et al.* (2015)

or 'often' make purchases online (2010 data). We can see here a clear pattern of peak internet activity among customers in the age brackets of late twenties and thirties. This profile stays fairly consistent whether online activity is defined as 'often' or 'sometimes' or other intermediate levels of engagement. It is interesting to see that there are now significant levels of e-retail adoption even among some of the more elderly customers, for example even in the 75–79 years age bracket more than 25 per cent of respondents have experienced some degree of online purchasing.

The additional dimension of gender is added in Figure 10.3. Here we can see evidence of another interesting trend, as it appears that among the highest adopters male customers are the most likely to be internet-friendly; whereas among other groups the bias towards females is more pronounced. Finally, in Figure 10.4 levels of adoption are shown disaggregated by household income. Again the relationship seems rather clear-cut as increasing levels of household income lead to progressively higher levels of e-retail uptake on all definitions.

The findings from the UK Acxiom data seem to be consistent with other studies on the demographics of internet retailing. For example, in the Dutch Multiscope study, which we introduced earlier, a pronounced skew towards online shopping was found in the 25–34 and 35–44 year age groups. Fifty-six per cent of the category 'frequent e-shopper' were found to be males, compared to only 30 per cent in the category 'non-e-shopper'. Although income is not included as a category in the Multiscope work, there is an equally definite bias on the basis of educational attainment which can probably be highly correlated with income. Respondents whose educational attainment is 'high' are three times more likely to be frequent e-shoppers than those whose educational attainment is 'low' (Farag *et al.* 2006). Similar trends were observed among online customers of the supermarket retailer Tesco as we noted in Section 10.2, with 80 per cent of customers in the 25–49 years demographic and 55 per cent males.

The point to emphasise here is that those demographic groups who are most likely to be high adopters of e-shopping are also most likely to be concentrated in the major metropolitan centres, and so this provides a partial explanation of why adoption rates can be higher in metropolitan areas. For example, in London the proportion of the population aged 25–44 was over 6.3 per cent according to the 2011 Census, compared to a national average of 4.4 per cent. At more localised scales the variation is more pronounced: for example among the 33 census wards of Leeds the share of 16–29 year olds (around 18–19 per cent for the city as a whole) exceeds 78 per cent in Headingley, 63 per cent in Hyde Park (both student areas) and 35 per cent in the City and Holbeck (a lot of new professional flats having been built there for young professionals).

A slightly different perspective on the question of customer profiling can be taken through the use of geodemographic categories (see Chapter 4 for a more detailed discussion of the method). The patterns in this analysis are muted to some degree in view of the averaging which is taking place between different population sub-groups.[1] In areas with a more elderly demographic (e.g. 'aged coastal resorts'), low levels of affluence (e.g. 'industrial legacy'), and low population density (e.g. 'agricultural fringe') then e-shopping levels are all relatively restricted.

The complementary perspective of internet provision is presented within Table 10.1, in which it has been possible to track the penetration of internet services across a period of seven consecutive calendar years of the ONS classification. At the start of the period, 2004, internet provision was already quite widespread among more than 50 per cent of households. However, the spatial variations mirror levels of utilisation quite closely. Low levels in particular were found in rural, coastal and deprived areas. By 2010, internet access had increased steadily but spatial variations had flattened quite noticeably. To some extent these trends might reflect the much more widespread availability of broadband internet connections at the end of 2010, whereas seven years earlier at the start of 2004 much more proactive steps were necessary for users to access internet services. Indeed the further analysis of Table 10.1 illustrates that although broadband connections were still more numerous in urban areas, the level of service penetration was actually highest in rural locations. From an infrastructure perspective it appears that the innovation-diffusion process was now effectively complete (see Figure 10.5).

The relationship between access, uptake and frequency of internet use is nuanced and complex. Farag *et al.* (2006) examine their Dutch internet data in

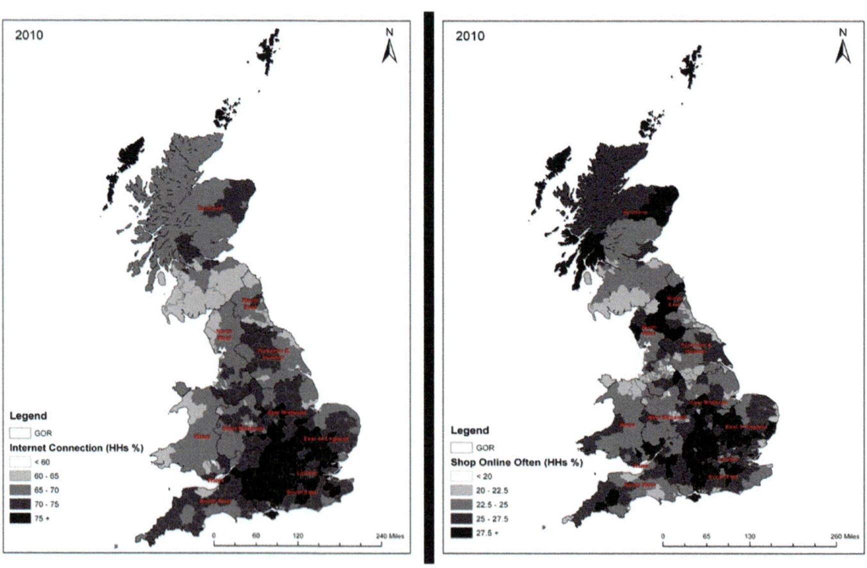

Figure 10.5 Current patterns of e-retail access and utilisation in the UK.
Source: Clarke *et al.* (2015)

Table 10.1 Internet provision by area type over time in per cent.

Code	*Name*	*2004*	*2005*	*2006*	*2007*	*2008*	*2009*	*2010*
A1a	Industrial legacy	45.50	47.04	49.95	52.46	57.84	62.62	66.69
A2a	Struggling urban manufacturing	41.44	42.89	46.34	49.26	55.22	59.90	64.04
A2b	Regional centres	44.33	45.67	49.26	52.03	57.83	62.24	66.38
A2c	Multicultural England	46.90	48.50	51.50	53.96	59.52	64.33	68.56
A2d	M8 corridor	45.88	47.53	50.41	52.63	57.61	62.43	66.54
A3a	Redeveloping urban centres	49.76	50.91	53.59	55.61	61.07	65.74	70.12
A3b	Young multicultural	55.46	55.99	57.79	58.75	63.90	68.23	72.91
B1a	Rural extremes	51.91	52.52	54.03	55.33	60.15	64.28	68.22
B1b	Agricultural fringe	54.58	55.18	56.57	57.69	62.45	66.92	71.11
B1c	Rural fringe	58.33	58.84	59.82	60.64	65.20	69.77	74.00
B2a	Coastal resorts	50.87	51.63	53.84	55.75	61.11	65.76	69.97
B2b	Aged coastal extremities	49.29	50.27	52.04	54.00	59.13	64.24	68.30
B2c	Aged coastal resorts	50.47	50.94	52.27	53.53	58.34	62.84	66.90
B3a	Mixed urban	53.05	54.01	55.96	57.58	62.55	67.24	71.41
B3b	Typical towns	54.53	55.88	58.14	59.62	64.42	69.06	73.43
B4a	Isles of Scilly	56.11	51.46	51.35	50.67	61.57	67.37	76.35
C1a	Historic cities	58.61	58.95	59.96	60.49	65.23	69.84	74.50
C1b	Thriving outer London	61.56	62.22	63.82	64.12	68.81	72.83	77.68
C2a	The commuter belt	63.63	63.75	64.58	64.70	69.17	73.24	77.80
D1a	Multicultural outer London	60.12	60.83	61.94	62.33	66.82	71.22	75.87
D2a	Central London	59.00	59.67	60.94	60.98	65.83	70.18	75.38
D2b	City of London	65.39	63.48	60.85	57.77	63.16	69.84	77.63
D3a	Afro-Caribbean ethnic Borough	55.06	56.20	58.58	60.01	65.20	69.68	74.29
D3b	Multicultural inner London	56.98	57.85	59.23	60.38	65.49	70.51	75.10
Total	**Total**	**52.92**	**53.87**	**55.89**	**57.38**	**62.41**	**66.97**	**71.28**

Source: Authors

relation to searching, buying and frequency of buying online for different demographics and location types. For demographics, the evidence endorses previous messages regarding higher propensity to search and buy online among young male customers. Education and income also appear to show a positive correlation, although on this sample size the inferences are not statistically significant. Familiarity with the technology, proxied here by the length of internet experience among the respondents, associates positively across all categories. In relation to location, it appears that urbanisation is positively associated with online buying; but when access to retail services is considered, this relationship is reversed for purchasing frequency. A plausible interpretation of these patterns is that city dwellers may be pre-disposed to access online services, but among more remote communities those that are able to make the transition are more likely to become heavily dependent on the technology for routine or repeat use.

10.4 Spatial interaction

The relationship between urbanisation and the uptake of e-retail services is a subtle one. This is evident not just at a national scale, but also at the sub-regional and the metropolitan level. For example, Figure 10.6 shows the online purchase frequency for groceries for Mid-Level Super Output Areas (MLSOA) in Yorkshire and Humberside based on estimates made by Kirby-Hawkins *et al.* (2017). MLSOAs are zones of roughly even population, hence the larger areas are typically those with the lowest population density. We can see that certainly a number of the larger areas have the darkest shading and hence the highest online purchasing frequency. However the pattern is by no means uniform.

A similar analysis, with a slightly more refined unit of geography, is shown in Figure 10.7, in which we can see not just the variation in uptake but also the locations of major supermarkets in the grocery retail network. The maps are built

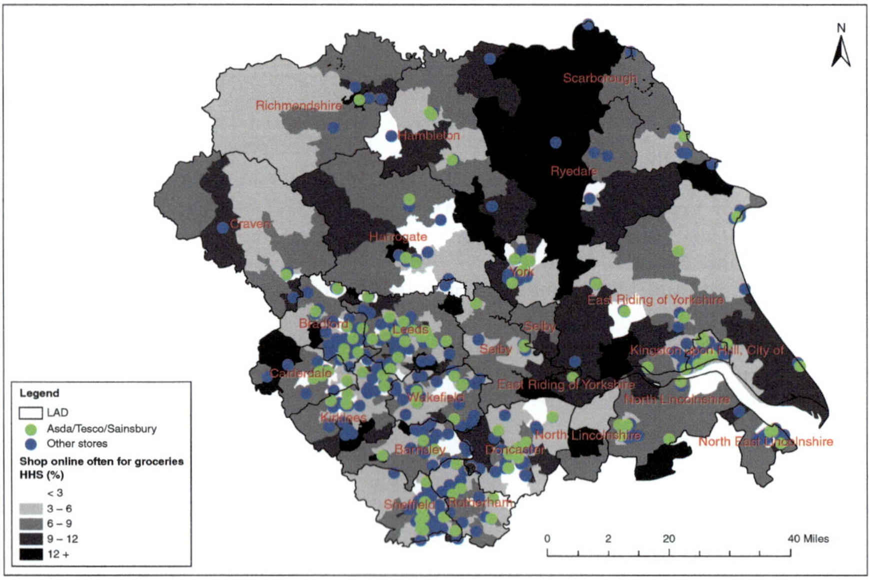

Figure 10.6 Estimating e-commerce demand for groceries in Yorkshire and Humberside.

Source: Kirby-Hawkins *et al.* (2017)

from individual responses to the ROP questionnaire which geo-locates customers on the basis of their preferred channel (online versus shop) and their accessibility to the grocery retail network. Cross-tabulation of this information allows a slightly more rigorous analysis of behaviour patterns, as displayed in Table 10.2. Although definitive explanations are elusive, the trends appear to be robust and consistent – online customers typically have poorer access to supermarkets for all the major grocery retailers than their offline counterparts. At the level of individual customers the 'efficiency' hypothesis appears to be firmly established.

A further complication that we have not yet considered is the impact of other 'distribution channels' beyond store and internet. Historically, the most important alternative to outlet shopping has been by mail order, where in

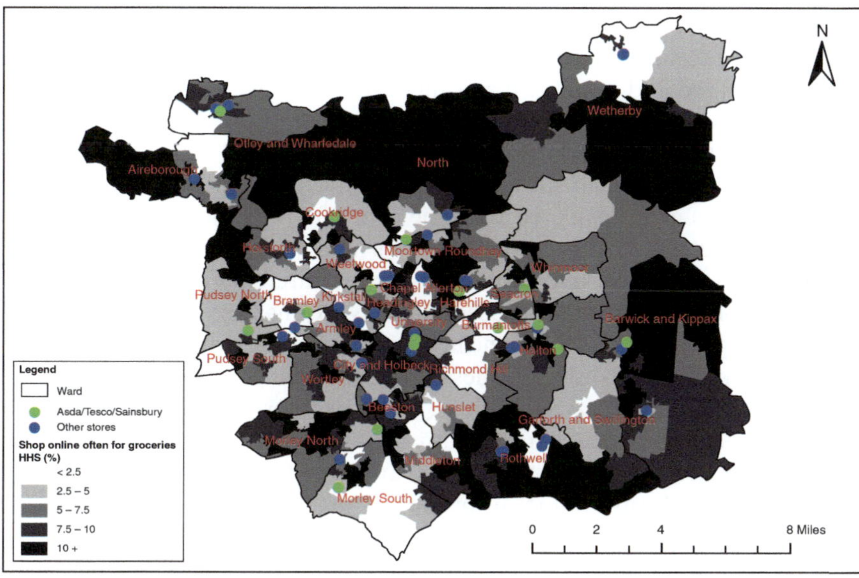

Figure 10.7 E-retail uptake in Leeds.
Source: Kirby-Hawkins *et al.* (2017)

Table 10.2 E-retail uptake and per cent market share

Asda			*Sainsbury's*			*Tesco*		
	Offline	*Online*		*Offline*	*Online*		*Offline*	*Online*
Min	0.06	0.04	Min	0.06	0.07	Min	0.02	0.10
Average	2.54	4.90	Average	2.44	3.14	Average	2.12	3.47
Median	1.69	2.54	Median	1.60	2.12	Median	1.36	2.15
Max	38.48	40.84	Max	30.42	32.14	Max	22.74	24.57

Source: Authors

the 1970s and 1980s companies such as Freemans and Grattan have thrived. There are certain markets in which mail order continues to be more influential than e-retailing: for example, in the case of clothing the latest evidence from the Acxiom poll shows that 11 per cent of mail customers exceeds approximately 8 per cent by internet, while for wines the shares are 4 per cent and 3 per cent respectively (G. Clarke *et al.* 2015). Purchasing direct from the manufacturer might be considered as a fourth distribution channel. It is ironic perhaps that in the past the most significant instance of purchasing direct from the manufacturer has been automotive retailing, albeit that customers have visited outlets specifically devoted to the sales of a specific retail brand (i.e. a franchise selling vehicles manufactured by Ford, Audi etc.). For many years the automotive industry has enjoyed a 'block exemption' from normal competition policies in order to sustain exclusive retail franchises (Brenkers and Verboven 2006). The irony being that automotive dealerships are now increasingly moving towards a multifranchised model in which competing brands can be compared and purchased at a single location, just as many other retail brands are becoming available direct from the manufacturer, via the internet, on an exclusive and non-competitive basis.

In addition to physical stores, e-retail, mail order and direct purchasing, online auctions are considered by some as a fifth 'channel' for distribution (Weltevreden 2007). Others consider this form of retailing as essentially C2C (customer-to-customer) and therefore slightly different from the core B2C (business-to-customer) retail process. Clearly in this case the retail interaction is strongly mediated through the internet, and this may also be true to a lesser extent of mail order businesses whose catalogues might typically be available in both a magazine/brochure and a web format. The general point to make is that in practice retail sales are mediated through increasingly complex networks of distribution channels, and that the relative value of each will tend to vary from product to product. Further analysis by Weltevreden (2007) found that even among e-aware consumers who have purchased by the internet, traditional retail channels continue to dominate the market for groceries (74 per cent of customer visits) and provide the leading channel for DIY (41 per cent), photographic products (38 per cent) and health and personal care (37 per cent), but more than half of all purchases of travel tickets (52 per cent) are mediated directly through the provider. On the other hand, virtual retailers have a dominant share for books (66 per cent), CDs (62 per cent) and especially optical products (74 per cent). Mail order is still the prime channel for erotica (62 per cent), furniture (64 per cent), underwear (77 per cent) and outer clothing (86 per cent); unsurprisingly perhaps 'collectibles' is by far the most important product for online auctions (92 per cent), but this is also the lead channel for toys (43 per cent) and jewellery (27 per cent).

Ultimately, consumer preference for alternative distribution channels may account for another finding in the research of Farag *et al.* (2006) – that the 'efficiency' effect is quite strong for products like clothing (in which the hold of the internet as a channel is relatively weak); but the 'innovation' effect is dominant for products such as travel tickets (in which the internet is a much more significant factor and in which the process of disintermediation is well-established). Thus in the case of clothing, the majority of e-retail customers were found to be in rural areas; for travel, the skew is towards urban areas. However, even more subtle processes are also at work in at least two regards. First, research has shown that while online purchase frequency is still highest in urban areas, the frequency of repeat purchases is higher in rural areas. In other words, getting people to change their habits and buy online is harder in rural areas; but once they have made the leap, they are more likely to maintain repeat activity on a regular basis. Second, there is evidence pointing to the use of multiple channels to support the purchasing process (Farag *et al.* 2006). For example, among people who have bought outer clothing online, 81 per cent used the internet as the information channel on which to base their decision, but also 42 per cent used a catalogue, 32 per cent had been to a city centre to conduct some form of physical inspection, and 8 per cent used intelligence gleaned from a circular or similar direct marketing material. The message is similar for online furniture sales with 89 per cent internet, 37 per cent catalogue, 33 per cent city centre, and 20 per cent circular. Supplementary information gathering across channels is an important consideration also for perfume and cosmetics, electronics, sporting goods, and even computers and software. The product sectors least affected by cross-channel comparisons are groceries, optical goods and travel tickets. These observations perhaps tend to reinforce the conclusion that online retailing is less a question of 'bricks or clicks' but much more 'bricks and clicks', as we shall argue again in the final section of this chapter.

As we shall see when we consider developments in 'big data' (Chapter 11) intelligence from market research sources such as ROP and Multiscope are increasingly being supplanted by transactional data from the retailers themselves. For example, many supermarkets and high street retailers now have loyalty cards which reward frequent customers and high spenders. From the point of view of the retailers, these cards have the benefit not just of promoting a higher 'share of wallet' but they also allow customer spending to be tracked and analysed. These data sets are potentially of value in academic as well as commercial research. The loyalty data are used to provide two measures of network performance for one example client – total weekly *sales* to customers in the region and *online* sales which are fulfilled through home deliveries. As background data, the sales area for both the client (*space*) and its competitors (*comp space*) is considered, along with the *population* and physical size (*area*) of each region.

Three derived indicators are calculated as shown in Table 10.3. *Urbanisation* is simply population divided by area as a measure of population density. Group provision (G prov) is the sales area per head of population for the retail client, and Competitor provision (C prov) is the equivalent measure for its retail competitors. Each indicator is also ranked from high to low.

The first test is on the relationship between e-business share and client provision. Among the top seven share by rank ('rank share', 1,2,3,4,4,4,4), we have six of the lowest level of client provision by rank ('rank G prov', 12,

Table 10.3 New indictors derived from loyalty card data.

Postal area	*Urbanisation*	*G prov*	*C prov*	*Share*	*Rank share*	*Rank G prov*	*Rank C prov*	*Rank urban*
BB	0.110692	0	0	6	11	12	15	6
BD	0.086471	0.005999	0.093306	8	4	10	4	8
DL	0.010015	0.008235	0.099106	7	8	8	2	14
DN	0.065823	0.010637	0.096968	6	11	7	3	9
HD	0.176416	0.017188	0.064613	7	8	3	12	4
HG	0.03083	0.023897	0.088287	8	4	1	6	12
HU	0.10384	0.008091	0.088251	8	4	9	7	7
HX	0.142953	0.01278	0.084286	6	11	4	10	5
LA	0.005786	0	0.033664	19	1	12	13	15
LS	0.181591	0.012166	0.09177	7	8	5	5	3
OL	0.059049	0	0.078612	16	2	12	11	10
S	0.211391	0.005978	0.08789	8	4	11	8	2
TS	0.030837	0	0.022612	10	3	12	14	11
WF	0.218564	0.011794	0.085394	6	11	6	9	1
YO	0.023582	0.017691	0.110491	5	15	2	1	13

Source: Authors

12, 12, 11, 10, 9, 1) – so apparently a clear substitution between physical and virtual channels.

A second test is a 'quadrant analysis' of client provision against competitor provision:

	C prov < 0.08	*C prov >0.08*
G prov <0.01	Rank 4 (share 12.75) [N = 4]	Rank 5 (share 7.75) [N = 4]
G prov >0.01	Rank 8 (share 7) [N = 1]	Rank 10 (share 6.3) [N = 6]

The hypothesis here would be that for given levels of client representation (G prov) then low levels of Competitor provision (C Prov) will tend to encourage higher levels of online use (because there are no alternatives). The evidence seems to support this idea (the shares fall as we move to the cells on the right) but the test is not conclusive, partly because there is only one observation in the bottom left cell (N=1) (although of course low N is an issue throughout this preliminary analysis).

A third test is between urbanisation and e-share. Here our expectation – based on the efficiency hypothesis – would be for greater uptake in rural areas. This appears to be borne out by the evidence once again, for example:

Average share in the six most urban areas = 7%

Average share in the six most rural areas = 12%

A final question concerns the interaction between client provision and rurality. For example it could be the case that provision is always lowest in the rural areas. Considering variation in market share against simultaneous deviations in both rurality and provision demonstrates that the highest e-shares are obtained in areas that are both rural and have low provision.

	S prov <0.01	*S prov >0.01*
Urban<0.1	12	6
Urban>0.1	7	6

This original piece of analysis using customer loyalty data therefore appears to support a number of intuitively appealing hypotheses about online behaviour (e.g. that it is promoted by weak physical store provision, especially in rural areas). Detailed work to add greater rigour to this investigation is currently underway (Kirby-Hawkins *et al.* 2017), for example:

1 There are a number of zeroes in client provision. This may be overcome through the adoption of more subtlety in the measurements – for example, the use of a provision ratio which accounts for cross-boundary flows.
2 A variety of statistical metrics could be applied, ideally involving some element of confidence or hypothesis testing, for example r-squared, Spearman's rank, or some kind of comparison of means for the quadrant comparisons.
3 The results can be drilled to a finer spatial scale, for example to postal sectors or districts which are up to one hundred times smaller than the postal areas adopted above – although measurement errors (e.g. in the computation of provision, or small number customer variations, may become problematic at these scales). The use of local authority districts as an intermediate scale could also be considered – there are about four times as many of these as there are postal areas.

To conclude this section it is useful to consider the implications of the three hypotheses above for store location research. First, there are obvious benefits for customer targeting and marketing through a greater understanding of the geodemographics of e-commerce users. In a general sense retailers would be advised to focus on locations containing more affluent, younger communities, while rural areas remain of high potential. More difficult to fully grasp at this stage is the relationship between face-to-face and online sales. If the relationships discovered for our client retailer in Yorkshire hold true across the country, adding a new store may at first allow some customers to swap back to face-to-face (the substitution effect). However, if sites can be found where new stores will increase the representation of a retailer in areas of low levels of competitor provision, we also forecast higher levels of online use for that retailer as brand awareness is higher and, with little competitor presence, the retailer might be able to enjoy high market share for both channels. A major research task is now to formally add these hypotheses/behaviours into the next generation of retail models

10.5 Click and collect

Perhaps one of the most interesting recent developments in e-retailing has been the massive growth in the integration of online purchasing combined with the physical collection of the product from a traditional retail outlet. This process is known as 'click and collect' and is growing as the preferred choice of purchasing non-food products for many consumers. Its success has been driven by a number of factors:

1 Consumers' dissatisfaction with home delivery. While supermarket retailers allow their customers a delivery window (typically between 1 and 3 hours) including evenings and weekends, the non-food sector offers a lot less flexibility. While supermarkets tend to own and control their distribution service most non-food online retailers outsource their deliveries to third parties, many of which have a poor record in what is known as the 'last mile' problem – successfully delivering the parcel to the customer. For products that require a customer to be at home (e.g. the parcel is too large to fit through the letter box) this can be particularly inconvenient and may either result in having to take time off work or having to visit a depot to pick up the parcel due to a failed delivery. Research by M. Clarke (2007) indicated that approximately 50 per cent of home deliveries of non-food orders fail first time.
2 Improved supply chain logistics. Providers of 'click and collect' services have a number of different fulfilment models but most offer next day collection. To achieve this most providers operate from dedicated distribution centres and outsource delivery to highly experienced distribution companies such as DHL and UPS. Customers are usually emailed when their order is available for collection.
3 Partnerships: Again there are a number of different models adopted by online retailers offering 'click and collect' services. John Lewis is a good example of one of these. John Lewis has only about 40 stores in the UK but through its sister company, Waitrose, it can provide over 300 collection points to its prospective customers. Perhaps the most interesting development has been the growth of third-party organisations that link online retailers with high street retailers. The biggest of these is Collect+ which has signed up 5,000 convenience and corner shops and in return for a small fee persuaded them to become collection points for a wide range of retailers. These include Boden, House of Fraser, ASOS, Amazon and eBay. Most of these collection points have long opening hours, often until 10 p.m., and some forecourt operators have 24-hour opening. Another key advantage of these outlets is that they offer a returns service for customers who either find they don't like the item they have ordered or, in the case of clothing, find it doesn't fit. For the store it also generates increased footfall, so customers picking up items may also spend money on products in the shop.

Click and collect is an interesting and fast growing merger of online and bricks-and-mortar retailing. Its popularity is likely to increase in the future. It raises fascinating questions for site location research. Retailers are now beginning to search for new geographies of distribution. Given the above discussion around delivery times etc., it is not surprising that consumers would like to collect their goods at a time and place that is more convenient. In grocery retailing, for example, the drive through a click and collect service point in the store car park can be more convenient than parking and queuing at an access point in the store. However, UK grocery retailers are beginning to get smarter in their thinking about convenience access points. Some have located access points (a better terminology perhaps than store) in industrial parks (e.g. Asda's click and collect kiosk in Green Park in Reading). In London, there is a spatial battle around serving customers who commute to work on the London underground. For example, Tesco now have click and collect facilities at Osterley, Newbury Park, Rayners Lane, Finchley Central, Arnos Grove and Cockfosters. In September 2014, eight new ones were planned around the network. At the same time (2014), Waitrose had facilities at six underground stations, all serviced by new temperature controlled lockers; Asda also had six (with six more in the pipeline); and Sainsbury's had seven underground locations.

The interesting geographical question is where else would make suitable click and collect locations? In addition to London stations, we saw in Chapter 2 how other transport hubs, including airports, were becoming popular. Asda have made no secret of their liking for petrol stations to be future hubs of both convenience and click and collect. Other possibilities might be schools, sports clubs, libraries or pubs (especially in rural areas). Thus the retailers are effectively creating new layers of retail geographies here – one that in the future it is going to be interesting to watch develop.

10.6 Reflections, conclusions and future directions

This chapter has attempted to elucidate and untangle some of the complex relationships in the electronic retailing marketplace. Among other things it has clearly demonstrated that evolution is still taking place at quite a rapid pace in relation to supply, demand and their (spatial) interrelationship. One level of this variation is at the level of individual products. In certain niche markets, such as travel tickets and collectibles, the uptake of e-retail has been rapid and is almost complete. For others, including groceries, the impact has been considerable, but by far the greatest share of activity remains oriented to traditional outlet shopping. On the other hand it can be argued that in such cases the potential for future transformation is still great, and it is in these situations that markets are still at their most dynamic.

Supply chains have been disintermediated to a more limited extent than was envisaged by many early commentators, particularly around the time of the millennium when the 'dot com bubble' affected many business and financial markets. This process has been most advanced for certain classes of high-value, single basket products (e.g. computer hardware) and for homogeneous, non-perishable commodities such as books and CDs. In many cases, markets have become quickly re-intermediated by comparison sites which could be viewed as a virtual counterpart to the high street in aggregating products from multiple virtual locations, with the added benefit of adding powerful search and (price) optimisation capabilities. Perhaps even more significant is the evolution of virtual intermediaries not just as an alternative but complementary channel of distribution. Here internet has now overtaken other ex-outlet channels such as mail order or leaflet promotions in significance across nearly all markets. This leads to some complicated patterns of purchasing behaviour which are as yet little studied or understood. The capability for further analysis is obvious. For example, Wilson-Jeanselme and Reynolds (2006) have begun to consider the relationship between demographics and channel preferences including customer preferences and attitudes to cost, quality and convenience. This work attempts to link demand and supply by suggesting that retail strategies are intimately linked to customer attributes and attitudes. For example, if the customer base is strongly focused on the quality of the product, internet retailers may need to pay special attention to the range and diversity of their stocks. If the issue is cost then maybe the distributional efficiency of logistics networks is somewhat more important.

In practice, quality and cost may be linked to demographic characteristics; for example, that younger, more style conscious customers are heavily focused on quality, but those of a low socio-economic status are driven by costs. Certainly we have seen that there are clear demographic patterns to e-retail uptake. There is obviously substantial scope for further demographic profiling of e-retail markets according to both a richer mix of socio-demographic characteristics and behavioural profiles (e.g. Longley *et al.* 2008).

The complicated mix of attitudes, behaviours, demographic and channel preferences has given rise to some intricate geographical patterns. It is still far from clear whether efficiency or innovation is the major process in structuring retail markets, although there does seem some evidence in at least some cases for a diffusion – from urban towards rural – which could be partially driven by service availability (i.e. broadband). We have seen that some of these changes are determined by the attitudes of suppliers as well as customers, and is some cases that the conservatism of established retailers may have slowed the pace of change (Nicholls and Watson 2005).

A richer perspective here might consider the impact of e-business (perhaps in its broadest sense) across the entire supply chain which embraces manufacturers, retailers, customers and their mutual interactions. In a fairly early study of this problem, Sparks and Burt (2003) have looked at a combination not just of financial impacts but changing relative power in the retail supply chain. At the manufacturing end of the chain it can be seen that electronic sourcing allows major producers to benefit from broadening the range of potential suppliers, extending their geographical search, and also faster and more efficient sourcing. In the same way, major retailers may be able to reduce costs by increasing their hold over suppliers who lack scale and market access. In contrast, however, it is the consumers who have typically gained most from the transition to virtual markets. This can be seen quite clearly in the case of grocery supermarkets, who are now expected to bundle and distribute products on their customers' behalf for zero charge, or at best for a consideration that is much less than the associated cost. In simple terms, whereas the cost of travel to the shops, product selection and transfer (to the home) have all historically been borne cheerfully by the customers, where e-business at least is concerned these costs have simply shifted to the retailer and are potentially very substantial. As yet, most retailers are not able to offset these increasing costs through supply chain efficiencies such as holding stock in warehouses or distribution centres rather than retail outlets, although this is a situation that is changing as the retailers begin to adapt their supply chains to changing market conditions.

Technological change also seems certain to continue apace and to influence future market development in significant but inevitably uncertain ways. Increasingly the notion of e-commerce itself has been superseded, or at least heavily augmented, by notions such as m-commerce ('mobile (phone) commerce') and s-commerce ('social (network) commerce'). The notion of Search Engine Marketing (now well-established as a major industry) has developed around the ideas of click-through advertising and the 'optimisation' of internet search engines to deliver predictable and desirable results from the perspective of individual corporations (e.g. if a customer is searching for sports shoes, how do I as a retail organisation ensure that my product appears first, or with the largest and most attractive advert). There is more ambivalence towards sophisticated data mining applications which could potentially unlock the value of social media. For example, at the time of writing, there is still a lack of evidence to demonstrate that the 'likes', preferences, attitudes and demographics of social media users can be effectively mined to promote appropriate products and services (although further evidence of potential developments in this area are considered in Chapter 11). On the other hand, it does appear that retail businesses will have to continue to integrate social media platforms as a means to retain market position. For example, in 2012 transactions using smart phones accounted for 13 per cent of online sales by value in the grocery sector. In 2015 it was estimated that 60–70 per cent of consumers would use their mobile phone for at least some part of the shopping process (searching and buying) (MobileShopTalk 2015).

In short, e-retailing is a dynamic sector with profound social effects for producers, retailers and customers. The impacts of technology have not yet stabilised, and in some sense may never do so. Further evolution is still expected of course, and will probably occur at an ever-increasing pace. There are fewer reasons to expect, however, that the importance of geography as a mediator of the retail process will decline, still less collapse, any time in the foreseeable future for all the reasons that we have endeavoured to express in this chapter.

Note

1 For example, it is obviously the case that 'aged coastal resorts' do contain families and young people; there will be patches of deprivation in 'Thriving Outer London'; and so on. In the language of Chapter 4, the spatial aggregation effects associated with the ecological fallacy become ever more significant at coarser levels of geographical resolution.

11 Big data analytics and retail location planning

11.1 Introduction

This chapter will consider the impact of recent technological developments on the ways that data can be accessed and analysed for the purpose of business planning in general, and retail market and performance evaluation more specifically. These developments have been variously characterised as 'big data', 'data science', and 'data analytics' as well as connecting to agendas such as 'smart cities'. Here we will adopt the compound term big data analytics, through which we hope to convey the message that it is not just new and potentially massive data sources that are of importance but the way that they are manipulated in order to inform theoretical debates or to support practical decisions. In Section 11.2 we will introduce big data specifically, starting with a definition of the concept and some simple illustrations to fix ideas. The scope and significance of efforts to mobilise access to new data sources for the benefit of social sciences in the UK will be considered, together with the implications for retail geography in particular. In the final part of the chapter some ethical difficulties and other restrictions will be evaluated.

11.2 Big data

In recent years, the use of the term big data has become increasingly widespread with reference to sources that are extensive in relation to their volume, variety and velocity (Laney 2001). There is an obvious trade-off here between economy of expression and nuance. In particular, it is notable that the first property of volume is easily overemphasised. It could be argued indeed that researchers have quickly become overly concerned with the size of the data resources at their disposal, when the velocity at which data are created (and often discarded) is also important. More significant than either of these we contend here is the variety of contemporary sources, providing a panacea of resources relating to people, attitudes, behaviours, corporations, capital flows, movement patterns, structure and change (to name but a few of the most obvious dimensions). In addition to the 'three

Vs' of variety, velocity and volume, a number of commentators have considered that reliability or veracity of big data should be considered as an additional key attribute (e.g. Mattman 2013) while additional Vs, including variability, visualisation, and perhaps most importantly value, have also been highlighted (Data Economy 2014).

To put some shape around this discussion, let's think about the Census of Population and Households. In view of our own roots, we will focus this discussion on the situation within the UK, but the same considerations will apply in the vast majority, if not all, countries where intelligence of this kind is collected on a routine basis. Census data provides extensive and high-quality coverage of core demographic variables, from age and gender to ethnicity and marital status. It contains valuable information about household structure and relationships. The data is somewhat less convincing with regard to social and economic attributes – there is coverage of educational attainment and occupation, but the important dimensions of income, wealth and expenditure are all neglected. Coverage of population movements and activity patterns is rather weak. Although it is possible to capture regular workplace interactions (the Special Workplace Statistics) and sporadic change of residence (the Special Migration Statistics) these are both incomplete and obsolete to some degree. Other patterns of interaction – with retail outlets, educational establishments, leisure facilities and other social and recreational opportunities – are omitted. Migration is considered in terms of a change of residential location between two periods of time, but the possibility of multiple moves is not incorporated. In both cases the data is a snapshot of a single point in time (i.e. census day; most recently 27 March 2011) so that weekly or seasonal fluctuations are out of scope, and since the Census is only conducted once every ten years, even with instantaneous publication, on average the information will be five years old, in which time many people may have changed jobs or moved again. Attitudes and lifestyles are almost completely without representation, on a full range of issues from political views and environmental sensitivities to favourite hobbies and membership of associations or clubs.

In contrast, big data can be thought of as a universe of available data about society and its constituent individuals. This could include commercial data which is captured through the business processes of organisations such as retailers, banks or (mobile) telephone/ digital media/ internet service providers; movements tracked through roadside sensors or smart ticketing systems; opinions collected via large-scale surveys or social media; as well as a raft of government intelligence from housing benefits to health records. Such data sets are often diverse in their content, have extensive coverage, and are refreshed continually, perhaps even generated in real time. In the next sections, examples of the exploitation of big data sources will be offered in both retailing and the social sciences at large. Potential issues in access and exploitation will be considered last of all.

11.3 Big data for retail planning

This section begins with an overview and introduction to the exploitation of big data for retail planning. Three examples will then be presented of ways in which new data of this kind has been applied in academic work. In light of these examples, the prospects and potential for big data analytics in retail research and in the social sciences at large will be highlighted.

11.3.1 Overview

As noted in previous chapters, spatial models and the associated analytical methods have been adopted in the past as a means for structuring relatively sparse data about retail systems. Indeed one view of models is that they provide a useful means for generalisation or the estimation of 'missing data'; for example, by interpolating detailed patterns of spatial interaction from known aggregates such as average travel distances or destination totals (e.g. Roy and Thill 2004). The advent of large-scale commercial data sets transforms this landscape quite fundamentally, but any suggestion that the analytic challenge is suddenly straightforward are wide of the mark. Indeed it can be argued with justification that deep approaches through modelling, simulation and theory are more important than ever if the descent from 'big data to bad science' is to be avoided (Siegfried 2013).

'Big data' is widely regarded as a radical development in the second decade of the twenty-first century. For example, the report by the Centre for Economic and Business Research on behalf of the SAS Institute (CEBR 2012) anticipates a rapid and profound change in the economy of the UK based on data analytics. The report of the UK government task force on administrative data (ESRC 2012) led directly to the formation of a Big Data Network in UK Social Science, with more than £60 million allocated to the Economic and Social Research Council (ESRC) for this purpose in the Chancellor's budget of 2013. Nevertheless, retail organisations have been actively exploiting data from customer transactions for many years already. For example, Humby and Hunt (2003) provide an enlightening overview of the methods used by Tesco to attract full value from its Clubcard loyalty programme in understanding and influencing the behaviours of its customers. From a vision created by Chief Executive Officer Sir Terry Leahy – 'what creates loyalty is how much we understand your life and what we do about it that helps your life' – the team at Dunnhumby created a customer segmentation bringing together data for 45,000 product lines and 10 million customers. In contrast to established segmentations such as geodemographics (see Chapter 4) the approach is distinguished through its basis 'entirely around what people did, not who they were'.

The 'Rolling Ball' methodology which was developed by Dunnhumby on behalf of Tesco starts with an intuitive process in which every one of the 45,000 products is assigned a profile on each of 20 scales (e.g. in relation to fat content, price, package size and convenience). The analysis starts with a small set of products that definitely have a perceived quality that is sought (e.g. 'adventurous products' might include extra virgin olive oil and ingredients for exotic curries). Customers with a high propensity to buy these products are identified, and then more products favoured by these consumers can be isolated. The set of adventurous products can be 'built out' until another product set starts to dominate. When launching the Tesco 'Finest' range, targeting by customer demographics (who they are) was not successful; but targeting by Clubcard segments (what they did) brought the desired result:

> The analysis showed that one segment of regular, loyal customers regularly shopped in 12 of 16 Tesco store departments. If each of these customers could be encouraged to shop in the other four just once every three months, Tesco calculated that the additional revenue could be worth £1.8 billion.
>
> (Humby and Hunt 2003; Humby *et al.* 2008)

We now review three contemporary case studies in which new data sources have been exploited in the context of academic research.

11.3.2 Example 1: Store performance and customer behaviour in seasonal markets

In a recent example, academic research addressed the difficult challenge of appraising visitor demand for supermarket retailers. The work combines store loyalty card transactions with directory sources to provide an improved understanding of patterns and opportunities (Newing *et al.* 2013b: see also Chapter 6). The aims and objectives of this study were to improve the accuracy of turnover models through incorporation of visitor demand; to develop estimates of seasonal variations in expenditure; and to assess the impact of geographical factors in local markets, accessibility, store quality and provision.

Customer transaction data was provided for 52 weeks in the financial year 2010/11. Individual transactions for a selected destination store in the county of Cornwall were aggregated for three different weeks during the year, for all local authority districts in England, as shown in Figure 11.1. The catchment area for a selected store, located in a major coastal resort, expands rapidly across the UK from the winter months into the summer and autumn.

Intelligence relating to visitor numbers, behaviour patterns, accommodation and hotel rooms was assembled from a variety of commercial reports, tourism statistics and trade bodies. A spatial interaction model (SIM) of retail expenditure (as described in Chapter 5) was developed to estimate the flows of income to stores from each output area. Destination characteristics, the demographics of residential areas, and relative accessibility are all key elements of the model. Through the disaggregation of residential demand by different seasons it is possible to model fluctuations of turnover in space

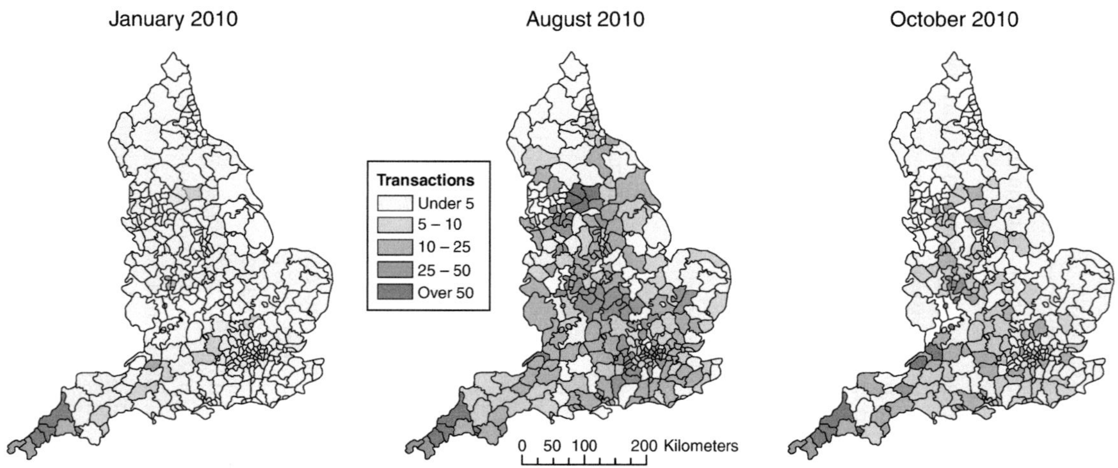

Figure 11.1 Expenditure variations by season. Number of recorded loyalty card transactions by district for a store within a Cornish coastal resort, UK.

Source: Newing *et al.* (2013a)

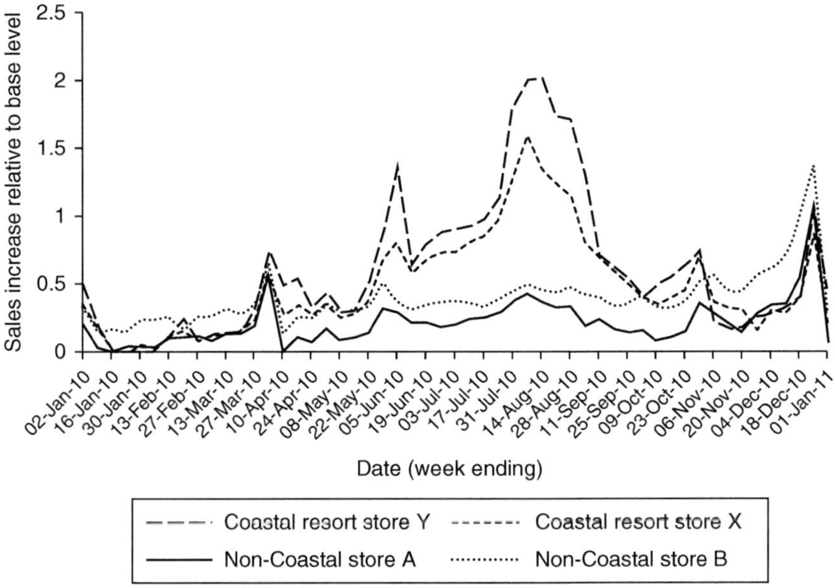

Figure 11.2 Weekly sales variation in Cornish stores.
Source: Newing *et al.* (2013a)

and time. The results demonstrate very significant variations in the volume of trade between different periods in the year, with coastal stores benefitting clearly from an uplift in trade in the summer months (Figure 11.2). Levels of visitor expenditure were found to be particularly difficult to estimate at the start of the summer season, where sales build-up is strongly linked to highly unpredictable weather patterns

In relation to the objective of providing improved predictive models of store performance, the research found that inclusion of visitor demand considerably improves revenue predictions at all stores. When residential demand is used as a driver for the SIMs, the estimates are robust in the middle of winter (January), but produce persistent revenue shortfalls at other times (August, October). After the inclusion of visitor spend, revenue predictions are substantially more accurate (Table 11.1) and within 10 per cent in most cases, although still subject to overestimation at stores such as 'Coastal Resort Store B' in which visitor expenditure here is partly driven by second home ownership with complex patterns of usage, and where day-trip visitors may also be a significant factor in driving sales uplift (see also the discussion in Chapter 6).

The example demonstrates the added value in customer interaction data from store loyalty cards for site location research. Spatial analysis has permitted not just the recognition of longitudinal fluctuations in shop catchments,

Table 11.1 Model performance by location and time of year.

Modelled store performance (August)	*Coastal resort store X*	*Coastal resort store Y*	*Control store A*	*Control store B*
Residential expenditure only	67%	72%	87%	87%
Residential and visitor expenditure	104%	107%	104%	99%
Modelled store performance (October)	*Coastal resort store X*	*Coastal resort store Y*	*Control store A*	*Control store B*
Residential expenditure only	46%	46%	77%	81%
Residential and visitor expenditure	91%	81%	99%	97%
Modelled store performance (January)	*Coastal resort store X*	*Coastal resort store Y Bude*	*Control store A*	*Control store B*
Residential expenditure only	92%	97%	98%	104%
Residential and visitor expenditure	95%	99%	100%	105%

Source: Newing (2013)

but also the evaluation of performance from model benchmarks with implications for merchandising, promotions and store formats. Similar methods could be applied for competitive analysis and regulation, as we shall see below. The findings of the Cornish grocery case study are suggestive of the importance of long-term fluctuations in population movement for retail planning and analysis. Similar experiments with massive customer surveys have also shed light on permanent moves (i.e. migration) and its impact on local housing and labour markets, for example in the work of Thomas *et al.* (2013). The availability of new and rich data sources promises to deliver even more radical, valuable and transformational understanding of short-term movement patterns in cities and regions, and to allow these flows – within a daily or weekly rhythm – to connect not just to retail geography but many other forms of service utilisation.

As noted above, government sources such as censuses and surveys provide an occasional snapshot of movement patterns in the long term (e.g. census migration statistics) and short term (e.g. journey-to-work counts). However these are not continuous and typically they are not up-to-date either.

Commercial sources promise the opportunity for access to near real-time assessment of mobility across all spatial scales, while simultaneously providing insights to the behaviours and motivations that underpin trends in space and time.

11.3.3 Example 2: Daily movement patterns

The exploitation of data generated by social media has spawned a rapidly growing research field. Work within the Geospatial Data Analysis and Simulation (TALISMAN) node of the UK National Centre for Research Methods has explored the value of these data sets as a means to represent and understand the movement patterns of individuals according to purpose, location and activity throughout the day. As already noted above, this provides a level of insight which goes substantially beyond the capacity of the census and other routine surveys.

The aims and objectives of this work are to detect diurnal variations and patterns in urban mobility; to relate movement patterns to their underlying causes and behavioural drivers; to develop, refine and calibrate simulation models of daily movement, and to evaluate their robustness for the representation of general trends. Data has been captured from the Twitter social messaging service and one of the reasons behind the rapid growth of social media analytics is the ease with which individual messages (tweets) can be acquired for any location or time period. The published research which is reported here has concentrated on the Leeds region, for which sample data is currently being acquired at a rate of around one million messages per year. National data sets are now being maintained by a number of academic groups within the UK. The National Library of Congress in Washington, DC, has copies of the global Twitter corpus going back to at least 2008, although the extent to which this can be shared and for what purpose are currently unclear.

Individual Twitter messages in the Leeds corpus comprise a string of up to 140 characters. Each message is also time-stamped to the hour, minute and second, and for users with a device which is location-enabled the spatial co-ordinates for each message are also generated. A number of approaches have been used in the analysis of the data: methods for clustering the data are important as a basis for understanding the daily activity patterns of individuals. Over the period of data collection, the time, location and content of messages can be examined. Spatial analysis of the data can reveal clusters of activity which Malleson and Birkin (2013) refer to as *anchor points*. The dominant cluster will usually, but not necessarily, correspond to the user's home location. This can be verified through examination of the textual content of the messages, as tweets from home will typically contain frequent references to television programmes, family, domestic meals and so on. The same principle can be extended to other activities and locations for 'if a home location attaches to a particular lexicon of words … then there is a further possibility to profile each anchor point according to the words used there' (Malleson *et al.* 2014). This principle is illustrated in Figure 11.3 which shows the spatial distribution of a selection of words that are frequently used in different contexts. Thus 'sleep' is an activity which is usually referred to in the home, whereas 'college' and 'pub' are normally both places nearby. On the other hand 'email' is a form of communication often associated with work, and this part of the daily routine is often reached through a railway or bus 'station' – both of these words are more likely to be observed at a suitable commuting distance away from the home. It seems equally likely that a user will refer to their 'iPhone' in any of these settings.

Clusters of Twitter messages and the associated anchor points can also be interrogated to understand the networks or activity spaces of the population. The hypothesis that residents of the city centre might be expected to move in more localised networks has revealed that 'spatial networks for individual users are typically more compact' (Birkin *et al.* 2014: 72), in the city centre, but that 'city dwellers do not have denser networks of activity and interaction, despite the higher concentration and greater access to opportunities in the urban core' (Birkin *et al.* 2014: 69).

Space-time clusters have also been explored within the context of geodemographic analysis. As we saw in Chapter 4, it is possible to characterise neighbourhoods according to the attributes of the people who live there. Through analysis of the Twitter data it is possible to extend this characterisation to the language adopted in different areas and to connect this to different activities and behaviours. An important outcome of this work is that it can illustrate variations in the geodemographic composition of neighbourhoods *in time as well as in space*. This idea has been illustrated by Birkin *et al.* (2013) who divided each day into four time slices and split the week into days of work and days of leisure. The occurrence of key marker words in each census ward

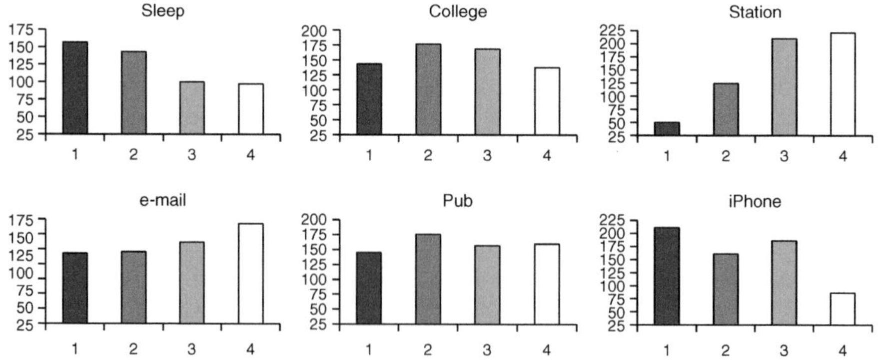

Figure 11.3 Concentration of selected words at four distance bands: 0–100 m; 100 m–2,000 m; 2–10 km; >10 km.

Source: Authors

was used to devise a six fold classification of areas, which was shown to vary through the week as well as across the city (see Table 11.2). Thus, in the city centre the language indicates a place of work in the daytime during the week, becoming a place of relaxation, entertainment and perhaps shopping in the evening, only becoming a place of quiet and tranquillity early in the morning. On the other hand, the suburbs are typically oases of domestic peace during the daytime, with a reorientation to sporting events and energetic leisure activities on weekend afternoons.

In order to express overall movement patterns and diurnal shifts in the population an agent-based simulation model was constructed by Birkin *et al.* (2013). Beginning with a spatial microsimulation model that represents the entire population of the city as a series of individual and household records, the Twitter data can be used to generate rules about the frequency and timing of transitions between activities, which typically also involve a change of location. This approach has many potential applications, for example, in being able to track populations of customers for retailing or other services throughout the day. Because the simulation is grounded in a complete synthetic representation of the population the overall results are less susceptible to bias than social media analyses which draw directly from a highly skewed and self-selecting sample of technology adopters. This concept was extended in a logical and powerful way by Lovelace *et al.* (2014) who combined the detection of both text and location to infer patterns of spatial interaction for museum visitors. The authors suggest that calibration of a SIM is an effective means to exploit social media data in relation to a wide variety of trip types and purposes.

Notwithstanding the great potential of emerging applications, the drawbacks in utilising Twitter and other social media data sets are obvious and significant. To date, the major adopters of the new technology have been a relatively limited group with clear biases, for example among the young and well educated (see also Longley *et al.* 2015). In particular, the propensity of a rather small cohort of individuals to generate the vast majority of messages

Table 11.2 Variations in area classification by message content.

	City *City centre*	*Horsforth* *Family suburb*	*Wetherby* *Retirement town*
Weekday 12 p.m. – 6 a.m.	Sweet Dreams	Sweet Dreams	Sweet Dreams
Weekday 6 a.m. – 12 a.m.	Daily Grind	Sweet Dreams	Family Life
Weekday 12 a.m. – 6 p.m.	Daily Grind	Idle Chatter	Family Life
Weekday 6 p.m. – 12 p.m.	Rest & Relaxation	Idle Chatter	Family Life
Weekend 12 p.m. – 6 a.m.	Sweet Dreams	Sweet Dreams	Sweet Dreams
Weekend 6 a.m. – 12 a.m.	Sweet Dreams	Sweet Dreams	Idle Chatter
Weekend 12 a.m. – 6 p.m.	Rest & Relaxation	Match of the Day	Match of the Day
Weekend 6 p.m. – 12 p.m.	Rest & Relaxation	Idle Chatter	Idle Chatter

Source: Authors

must always be carefully moderated. One potential counter would be to extend studies to a much bigger population of mobile telephone customers.

Another interesting source of consumer data is that generated from mobile phones. These devices are connected to a network of masts, and as the device moves about, the mast that the phone is connected to changes. Every time this happens an 'event' is recorded and captured by the mobile phone operator. This builds into a huge data set on consumer behaviour in terms of their location and how long they spend at particular locations. Using this data, aggregated to ensure anonymity, it is possible to identify the number of people present in particular locations at different times of the day and different days of the week. Clearly, for a single mobile phone operator, their customers only represent a subset of the universe of consumers. However, it is possible to use weighting methods to convert this sample, even though they are very large samples, into a picture of the behaviour of all consumers. At Leeds we have been working closely with Telefonica Dynamic Insights. Telefonica in the UK own O_2, one of the four big mobile phone operators in the country. To give an idea of how much data they capture, Telefonica estimate their customers generate 1.6 billion 'events' each day from around 18 million individual device owners. So on average for each device owner we have 88 distinct locations each day where we know a relatively accurate geocode (the mast that the device is attached to) and what is known as the 'dwell time' – how long the device was attached to the mast. So, for example, if a device was attached to the same mast between 9 p.m. and 7 a.m. we could reasonably impute that was the device owner's normal place of residence. Similarly, if the data recorded a dwell time of greater than, say, six hours at the same mast during the day, we could reasonably assume that is the device owner's place of work. Exceptions will of course apply but in aggregate terms this data should provide interesting insights into consumer mobility patterns. To provide some examples of the potential power of this data we present two case studies: one from retailing, the second from transport.

In 2013 a major new retail development was opened in the city of Leeds, known as the Trinity Centre, which was planned to revitalise the area between the core retail area and the train station. Using Telefonica data we can calculate the number of individuals present in the centre along with those in the other main retail hubs in Leeds. This is shown in Figure 11.4. As can be seen there is a major uplift in the number of visitors on the day it opened, the weekend after it opened and a week later, which was Easter weekend. We can also examine the impact on other retail hubs in Leeds city centre.

The second example relates to consumer travel behaviour. Train operators in the UK have patchy information on how many people travel between stations but have no data on where passengers originate their journeys or where their final destination is after they have got off the train. Using mobile phone data it is possible to identify passengers on trains (via their attachments to masts along rail routes) and identify their residential origin and their final destination. Figure 11.5 and Figure 11.6 demonstrate this for journeys along

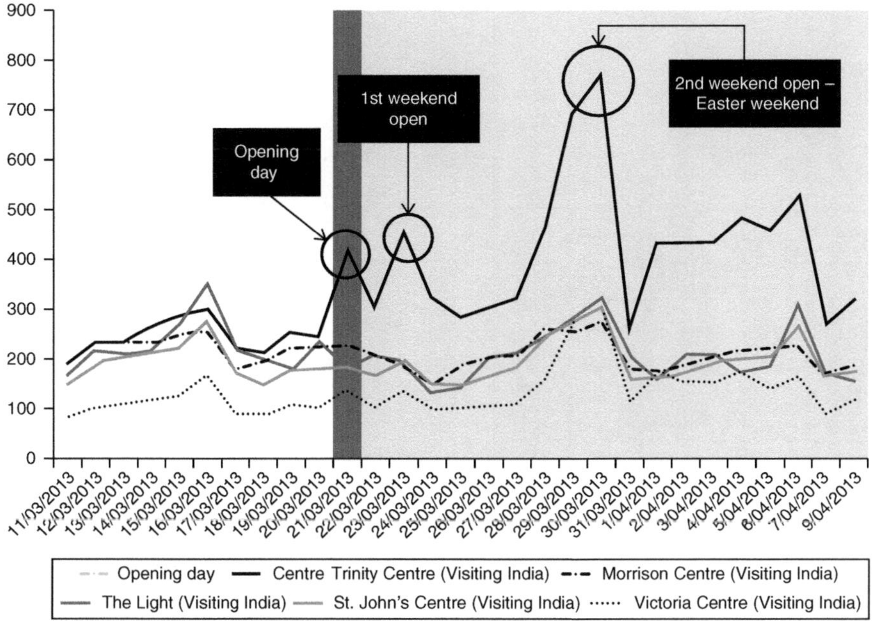

Figure 11.4 Plotting the number of individuals present in Trinity Centre Leeds.
Source: Telefonica Data Insights (2014)

the East Coast main line in the UK. Not only can we identify the numbers of passengers travelling between stations but also where they come from and where they end up. Figure 11.5 shows where people who boarded a train at York station originated from and also the origins of all passengers who used the East Coast main line on a particular day of the week in 2013. Figure 11.6 shows where customers who arrived at Kings Cross ended their journeys on that particular day. This type of information would be of significant value in analysing proposed transport plans (e.g. HS2) by having true origin – destination matrices rather than just station to station data. Also, it may help to understand movements of people to shopping centres – assisting us to better calibrate existing store location models and perhaps revise these to incorporate multipurpose trip-making (discussed also in Chapter 5).

11.3.4 Example 3: Market efficiency, customer equity and retail strategy

A third possible use for big data is in the assessment of retail strategy, market positioning and competitive analysis. Work of this type is potentially of substantial commercial value, but also provides insights into the effects of regulation and the efficiency and effectiveness of service provision in different business environments. Good examples of this type can be found among

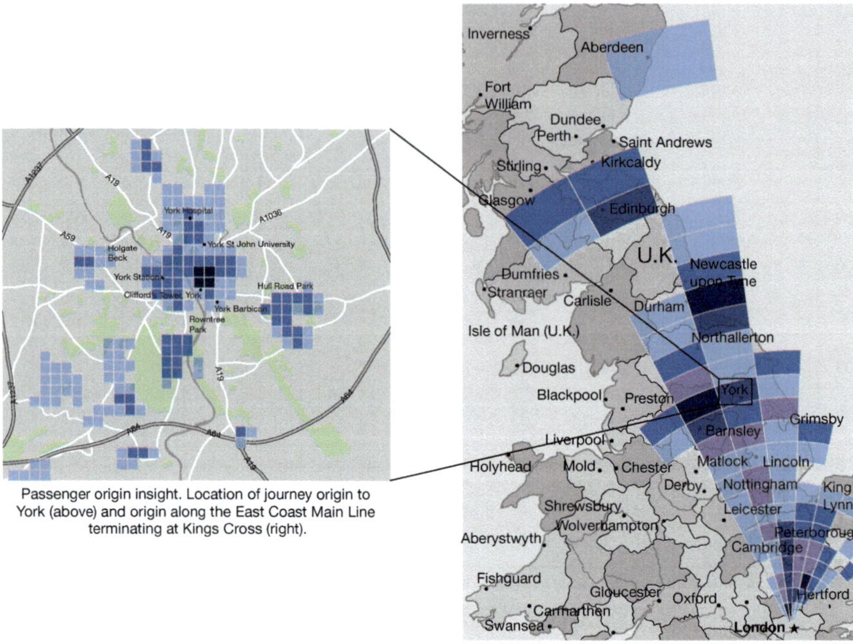

Figure 11.5 Plotting individuals via mobile phone usage on the UK East Coast train line.

Source: Telefonica Data Insights (2014)

utilities such as the water industry, which in the UK comprises a set of geographical monopolies, allowing households no choice in the provision of water (unlike other utilities). The supply of water to a property is mandatory and there is no contract with the occupant and no ability (enshrined in legislation) to cut a property's supply off. Supply is also highly regulated by agencies including Ofwat, the Drinking Water Inspectorate and the Environment Agency. Hence the industry needs to manage consumer debt carefully. Water might seem a strange example to use in this book – but of course water is a product that is in essence needs to be retailed just like groceries or furniture.

The problems of the water industry in relation to consumer debt are great (M. Clarke *et al.* 2012). In 2008 unrecovered debts cost the UK water industry £1,225 million, an increase of 48 per cent from 2005. The belief that debt levels are associated with income deprivation, with higher levels of deprivation associated to higher levels of debt, is widespread but unproven.

The aims and objectives of this case study were to provide improved evidence on debt collection within the water industry; thus to enable industry regulators, the UK government and other stakeholders to have access

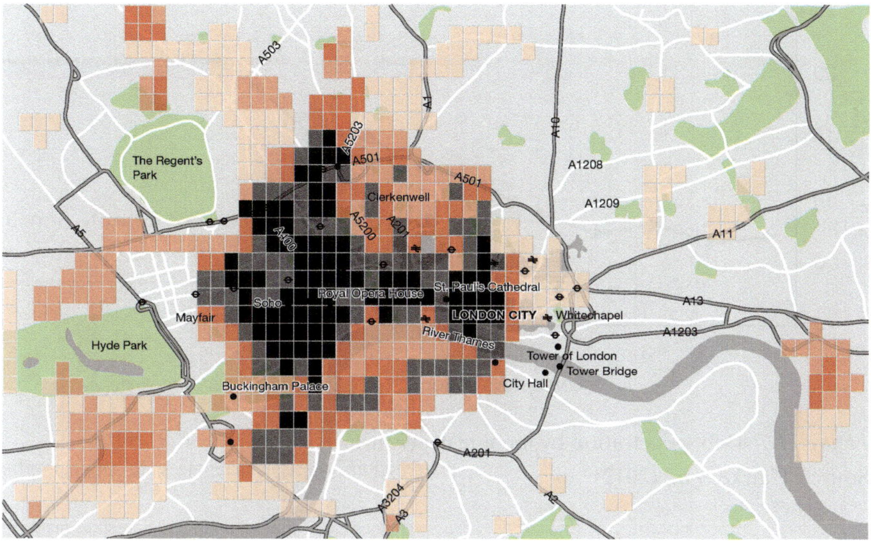

Figure 11.6 Passenger destination insight (terminating at Kings Cross).
Source: Telefonica Data Insights (2014)

to information which better informs the debate on future debt strategies. Specific objectives were to investigate the correlation between debt collection performance and income deprivation across the industry; to investigate outliers; and, where feasible, provide explanations for differences in relative performance.

The study used data relating to both consumer debt and deprivation. The collective supply area covers approximately 81 per cent of the population of England and Wales. Small areas were stratified (by income deprivation decile) and consumer debt was sampled at random for 5 per cent of households (almost 1 million properties) within each supply area. The UK Index of Multiple Deprivation (IMD) reflects different 'domains' which make up the overall IMD score. The domain components are Income, Environment, Health, Education, Access, Crime, Education and Housing. This data is available for Lower Super Output Areas (LSOA) at both an aggregate level and at the individual component level. The analysis included only domestic properties and focused on debts of greater than three months. Two companies were unable to disaggregate debt by duration. All analysis was anonymised to protect the privacy of individual householders. Within each water company sample it was possible to identify the level of debt penetration – the percentage of properties having some level of debt (more than three months old and

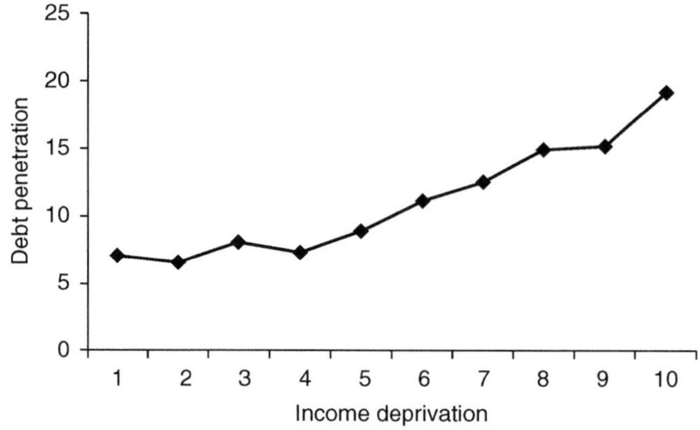

Figure 11.7 Debt penetration by income deprivation.
Source: Clarke *et al.* (2012)

greater than £10 – see Figure 11.7). Those properties subject to local authority collection agreements were excluded from the debt penetration calculation.

Debt intensity provides an alternative measure of indebtedness – the average debt associated with all indebted properties. All other things being equal, a high debt intensity should be associated with higher bills, but a high debt penetration is not necessarily associated with a high average debt intensity.

The findings of the investigation in relation to debt penetration revealed that where income deprivation is lowest, approximately 7–8 per cent of measured properties are indebted. Debt penetration increases to 20 per cent in those areas with the highest income deprivation score. For measured properties, debt intensity is £150 per property in the least income deprived areas, rising to £450 per property in the most deprived areas (see Figure 11.8).

A combination of consumer data from the water industry and 'open' government data on demographics and deprivation can inform debate about collection practices, pricing and policies for collection, enforcement or amelioration. It can provide valuable information to commercial organisations, regulators and consumers. The case study illustrates the willingness of water utilities to pool data, and the ability of spatial and demographic analysis to generate intelligence with policy value and social insight.

11.4 Discussion and implications

So far in this chapter we have described three substantial case studies in relation to long-term demographic movements, short-term demographic fluctuations, and the regulation and diagnosis of market failure. As the use of big data becomes more widespread in both commercial and academic practice we expect to see much more diversification of retail data analytics. We have seen

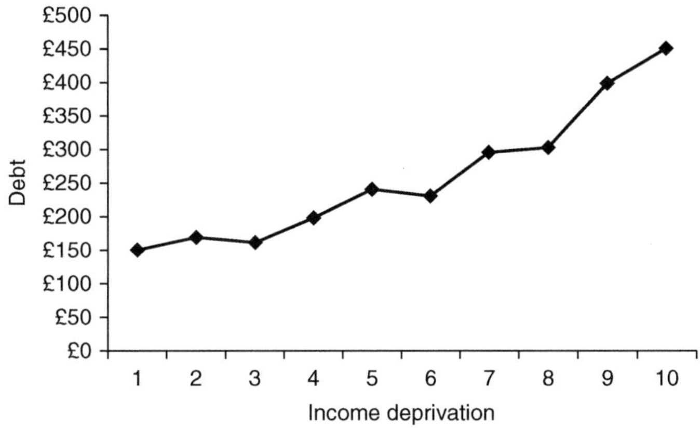

Figure 11.8 Debt intensity by income deprivation.
Source: Clarke *et al.* (2012)

in other chapters that many important insights can be achieved by working alongside retail organisations, but that the challenge of understanding markets and their customers is often hampered by access to good data of the right quality. If major retail organisations could be persuaded to share millions of customer transactions for both academic and business research, the range and quality of projects to be undertaken could be transformed quite dramatically.

For example, a major retailer could explore the impact of online shopping within various retail centres and market environments, and explore variations between different customer types, the actual and potential for behaviour change for different product categories, impact of emerging channels, such as click and collect, and the effect of changes in the local economy, such as economic regeneration, new housing or evolution in the strategy of a key competitor. It might also examine variability in daily consumption patterns in different locations, with a particular emphasis on convenience retailing (e.g. to investigate the impact of store layouts and stocking policy on performance) perhaps incorporating dramatic shifts in the profile of both city centre and suburban outlets between the day and night times. It could consider the overall demographic and social consequences of all of these trends towards new formats and distribution channels; for example, along the lines of a model of innovation adoption – which customers are the leaders, which are the laggards, and is there a momentum effect promoting more widespread penetration in particular geographical areas or community types? It might also explore how profiles of purchase frequency and basket size vary across different demographic segments or location types.

Cross-referencing purchasing information against the outlet location could provide valuable intelligence about the daily, weekly and seasonal mobility of customers, and perhaps spark revealing questions about the relationship between network density and brand loyalty (am I more likely to be a loyal or regular customer if there is a convenience store at the railway station where I work?). Store performance could be related to pedestrian flows from mobile sensors or other sources such as Wi-Fi traffic to understand the importance of micro-location and adjacencies. Changes of customer address could be used as a novel means to investigate migration, and as an early detection system for market entry opportunities in areas of rapid growth in an appropriate customer base. Variations in customer behaviour could be assessed by linking sales patterns for individuals or by stores to the presence of ethnic minorities or other socio-demographic concentrations (e.g. in the elderly or unemployed – cf. food deserts, see Chapter 5). Loyalty cards could be used to explore the interaction between access to stores and the use of online channels (as hypothesised in Chapter 10). Also, monitoring changing expenditure patterns could be important to evaluate the effectiveness of promotions in building market share, or as a defensive measure in relation to competitor activity such as the opening of a new discount store.

The potential for a better understanding of healthy lifestyles and food consumption is also very exciting. For example, the construction of indices relating to consumption of fruit and vegetables, alcohol and tobacco would provide new perspectives on the behaviour patterns of different socio-demographic groups and relationships between consumption and provision (important within food deserts again). Understanding customer frequencies and movement patterns could also provide improved evidence about the carbon footprint of customers. This could provide a fascinating interaction with other socially responsible preferences such as a propensity for locally sourced products or environmentally friendly packaging. Is it possible to aggregate customer behaviour patterns, for example, across the entire portfolio of trips, baskets and products in order to relate the overall impact to desires and intentions: that is, which customers are doing most to eliminate the 'attitude-behaviour gap' between their aspiration and real expenditures? (Papaoikonomou *et al.* 2011).

As some of these examples already hint at, one of the most exciting aspects of this agenda is that applications are by no means limited to retail geography and marketing. In relation to health care we suggested above that indicators of diet and nutrition might be one potential output. If these indicators were linked to clinical records relating to local regimes of treatment and morbidity, powerful new evidence could be revealed for both medicine and social care. An intriguing set of questions might be explored in relation to predictions in the short term as well as the long term. For example, on special occasions or events (New Year, Halloween, the Notting Hill Carnival or Leeds Music Festival) is it possible to forecast an increase in emergencies later in the evening from an excess in alcohol purchase earlier in the day? In relation to crime, the received wisdom in environmental criminology is embedded within concepts such as 'routine activity theory' (R. Clarke and Felson 1993) which could now be operationalised and tested in new ways. The use of smart ticketing within transport systems could be used to inform the frequency, pricing and capacity of services and infrastructure. The deeper implications of work such as this could run to inward investment and house-building, as in classical land-use and transport modelling. It could also lead to new domains, such as understanding the transmission of diseases like influenza (Ginsberg *et al.* 2009) or Ebola (Carter 2014).

A facet of these examples is that the data does not just inform established regimes of behaviour and interaction, whether in retailing or elsewhere, but the novelty of the data also often reflects a transformation in the underlying mechanisms for the conduct of commercial business transactions and the delivery of public or private services. While the shift from retailing 'across the counter' to online models of fulfilment is an obvious example, there are many other aspects to this process, both established and emergent. As a final stage in considering the impact and opportunities that could be presented by big data, a commentary on some of these possibilities is now provided. The ability to preserve large amounts of data about individual customers over long periods of time could allow for enhanced understanding of repeat behaviour, loyalty and customer value.

The advent of social media has implications for virtual as well as physical networks. Retail organisations will increasingly seek to influence these social networks; for example, by crowdsourcing market intelligence and through controlled viral marketing campaigns. The emergence of sophisticated forms of retail advertising and promotion can already be seen through cases such as 'Starcount', which seeks to exploit the importance of brand leaders (e.g. sportsmen and media personalities) in pushing consumer messages to their 'follower' networks.

Many commercial organisations are now employing 'sentiment analysis' to determine the attitudes of customers and their reaction to specific stimuli such as promotional messages or a product purchase. Sentiment analysis will typically involve the use of text recognition and data mining techniques to interpret the content of social messages such as tweets, blogs or status reports on social media. Monitoring the content of social media can allow companies to tailor communications to the attitude and mind-set of individual consumers. This approach advances on conventional techniques for direct marketing and customer management by allowing not just for qualifiers such as demographics (age, gender, affluence), or even through the incorporation of lifestyle and interests, but by context-specific criteria such as mood and circumstance. Similar techniques of sentiment analysis have been employed in other domains, such as the study of voter behaviour in elections, and in the appraisal of the London riots of 2011 (e.g. to demonstrate that the primary role of social media was a positive contribution to control and reconstruction, rather than a destructive influence to promote criminality – Baker 2012).

Much of the emerging data concerning individuals is geographically referenced. This will give new impetus to location-based services such as retail solutions and promotions driven by current activity; for example, alerting consumers to spending opportunities, identifying occasions and preferred timing of transactions.

While we have already seen the potential of big data to track movements across a variety of time horizons, these patterns also scale naturally to the national and international scale. Thus global movements of both tourists and business travellers can be tracked – this cohort is typically a group of high spenders with significant individual value, and whose opinions may have an important influence on global perceptions of brand and business performance.

Where commerce leads, public organisations and governments will increasingly adopt similar methods directly to maintain customer satisfaction, deliver services and monitor feedback. For example, health systems are increasingly used to provide advice and diagnostics for patients as a low-cost and universally accessible alternative to clinical practice. The control of online media and communications are seen more and more as a useful device to influence political activity; for example in turbulent regions and at times of conflict (Ben Moussa 2013).

11.5 Obstacles and limiting factors

Although there is potential for the exploitation of big data in retail research and related domains there are also a number of limiting factors. At the time of writing it is unclear how strong a brake these obstacles are likely to prove in restraining the development of novel applications. Here the obstacles will be considered using four broad headings of ethical, practical, methodological and educational.

11.5.1 Ethical concerns

The ethical and legal dangers from big data have been the subject of widespread discussion. In a rather well-known case study, analysts at the American retailer Target are credited with the development of predictive models for pregnancy that can seemingly outperform even the most intimate of 'local knowledge' (Duhigg 2012). In a notorious anecdote 'so good it sounds made up' (Hill 2012), the title of the article says it all: 'How Target figured out a teen girl was pregnant before her father did'.

A major concern in the exploitation of big data is whether informed consent has been provided for the way that data is being used. Informed consent is an established standard in the use of personal data as a means of protecting privacy and the personal ownership of an individual's data. In many of the cases presented in this chapter data has been provided in one context but the data owners or their nominees are looking to exploit the value of the data in another context entirely. According to Birkin *et al.* (2014), if this data is pooled and used for purposes other than those to which the participant has agreed, the data is being used without informed (or possibly any) consent.

The counter-argument that consent has been obtained through the signing of a contract by the customer permitting a wide range of secondary uses is far from watertight since the contracts themselves are often so voluminous that it is not reasonable to assume that the conditions have been fully absorbed and understood. A major outcome of the recent UK Parliamentary Select Committee investigation into social media data and real-time analytics has been a recommendation to simplify the terms and conditions of contracts, requesting that 'companies … commit(ting) themselves to explain to customers their plans to use personal data in clear, concise and simple terms' (House of Commons 2014).

The security of data is also a relevant issue under the ethical banner. Customer details are regularly misappropriated by retail organisations, often as a result of 'hacking'. Once again Target has been accused of that, having surrendered credit card details on 40 million customers, and telephone and address details for a further 70 million customers in an attack on December 2013 (Washington Post 2014). The fact that benefits from the analysis of data in a commercial context are typically entities other than the original data providers (i.e. the value accrues to the retailer or its agent rather than to the customers who provide the data), and that outcomes are not demonstrably in the public interest, is also a concern for many people. This appears to have been a major issue in a related controversy over UK medical records in 2014, where patient concerns about the disclosure of personal data to the insurance industry may have set back the cause of big data in the health sector by many months or years (Telegraph 2014).

11.5.2 Practical concerns

As we have noted previously, one of the reasons for the growing popularity in the use of Twitter for social media analytics is simply ease of access to the data. The same cannot be said for commercial data. Such limited applications as there have been typically rest on bipartite agreements between retail organisations and a university or research group. To address this challenge, the major government sponsor of social science research in the UK, the ESRC, has now established a Big Data Network including a group of Administrative Data Research Centres as well as four Business and Local Government Research Centres. Through the Big Data Network, a Consumer Data Research Centre (CDRC) will be established by means of coupling two programmes directed from the University of Leeds and University College London. A primary objective of the CDRC is to negotiate access to commercial big data sets for the whole academic community, and hence oversee the establishment and curation of 'consumer data' as a long-term resource. Building trust and raising awareness among the public with regard to the responsible use of data and the differentiation of commercial versus social interests will also form an important part of this programme.

11.5.3 Methodological concerns

The naive view that 'bigger must be better' does not apply to data. Many survey data sets are carefully weighted using processes of 'stratified sampling' to represent and balance different segments of the population. In comparison, big data sets are riddled with skews and biases. This is true of social media such as Twitter, but equally to mainstream retail organisations, which all have distinct profiles in their spatial distribution and the social mix of their customer base. Methodological innovation is therefore a key requirement to find ways in which the data can be used reliably. In one of our examples above it has been argued that simulation techniques are one possible way to achieve this, although many other options for reweighting consumer data samples are also available to 'careful' researchers. Without the means for the careful extraction of trends and patterns, however, big data becomes simply what Lovelace *et al.* (2014) refer to as 'big noise' (cf. Silver 2012).

It will also be important for both researchers and users to continue to recognise the limitations to the technology, especially regarding predictive analytics. Hence, although 'predictive policing' may be increasingly straightforward 'to select what streets, groups and individuals to subject to extra scrutiny, simply because an algorithm pointed to them as more likely to commit crime' (Mayer-Schonberger and Cukier 2013), such measures will do little to improve the efficiency or effectiveness of interventions unless they are based on robust foundations of understanding.

Emerging forms of data such as text and video will also demand the generation of new forms of information extraction. The example of sentiment analysis for the detection of underlying meaning and attitudes in text strings is one example that has already been touched upon earlier. Various machine learning approaches capable of tracing patterns in unstructured data are another class of methods that can be expected to grow in importance. In the future it could easily be that the pictures we post online, or the images we scan when searching for products, reveal more about preferences and behaviour patterns than the actual transactions, which are the current focus of consumer data analytics.

11.5.4 Educational concerns

Despite the considerable hype currently surrounding big data, concrete examples of its utility are still relatively rare. A major reason for this is a lack of awareness within the academic community, in corporate teams and among the general public. New programmes in big data analytics and data science are now beginning to appear in many higher education institutions, and these may be expected to build quickly in the coming years. In an ideal world this would be complemented by burgeoning programmes of professional education (e.g. through short courses or online learning). This is of particular importance to business and government organisations – if the Centre for Economics and Business Research is correct and 58,000 new data scientists are needed by

2017 then the 'production' of this resource is a major immediate challenge for education and training (CEBR 2012).

The most successful new courses in data science will very likely have two key ingredients. First, they will be multi-disciplinary in their emphasis. In order to solve the challenges and grasp the opportunities, the best research – whether in a corporate or academic context – will demand a combination of technical skill and domain understanding. It is unlikely that the big data world of the future will be ruled by computer scientists alone, nor by students of business or geography, but by a rich alliance of these and other disciplines. Second, they will be founded on strong partnerships between academia and commerce. As we have seen, the very minimum requirement for this interaction is a free-flowing exchange of data and the associated research. Direct investment as in the recently established KPMG Centre for Advanced Business Analytics at Imperial College London (TechCityNews 2014) may also become an increasingly important feature of the big data research landscape.

11.6 Conclusions

Big data has been identified as one of the 'great technologies', having the ability to power transformations throughout the economy in the first half of the twenty-first century (Willetts 2013). Retail analytics stands to be a major beneficiary from big data because of the volumes of information that are being generated through conventional retail channels, for example, with store loyalty cards, as well as by social media and other emerging channels of distribution. All forms of consumer interaction and service uptake will share in these benefits, from transport to health, and from housing to crime prevention.

In order to maximise the gains from emerging sources of data, a more innovative approach to the development of new methods for analysis may be required. Notwithstanding any commercial interests, this is one compelling reason that it will be necessary for businesses and government organisations to share their data more freely with the academic sector, and with other users including the public, in the future. A second reason is that analytic value is produced, not just by methods but by people. A shared approach to data-driven problem-solving is required to build an appropriate base of skills and education. The networks and institutions to facilitate appropriate partnerships between business, government and universities are already in place, for example in the shape of the ESRC Big Data Network mentioned above. An exciting future is promised in which social research, commercial planning and public policy are all more richly informed by real evidence than ever before. However, effective exploitation of the data will also require new models of informed consent and social acceptance of the importance of shared data in the common interest. In the short term, further work on the legal and ethical frameworks for big data analytics may be every bit as important as developments in computation and social science methods.

12 Conclusions

In this book we have tried to provide a contemporary account of the role that location analysis plays in multi-channel retailing. We have emphasised that despite the growth of new channels, especially e-commerce, over the last decade issues surrounding the location of consumers, the location of outlets and the use of different channels to link the two remain of considerable importance. To summarise we identify a number of themes that have recurred throughout previous chapters.

12.1 Changing retail dynamics

Retailing, in its widest sense, is a sector that has always been characterised by rapid change. Over the last decade many familiar brands have disappeared from the high street in the UK. Some of these (e.g. Woolworths) were the victims of changing consumer preferences and increased competition. Others, such as Blockbuster, were the victims of changing technology, in their case the ability to watch or download movies via broadband services or online competitors such as Netflix. On the other hand, a new wave of retailers has emerged or has rapidly expanded. Of these some have taken advantage of increased consumer concern for value for money during the prolonged recession. Poundworld and Poundshop are two UK retailers whose outlets sell everything from food, clothing, and health and beauty products for a pound. Greggs, the UK high street bakery store, has expanded from 240 stores in 1984 to over 1,600 stores in 2015, based largely on a highly aggressive pricing policy. Supermarkets are also changing rapidly. The dominance of the big four UK brands (Tesco, Morrisons, Sainsbury's and Asda) has been threatened by the expansion of Aldi and Lidl who have expanded very rapidly over the last decade but have ambitious plans to continue doing so over the next few years as well. Both Tesco and Morrisons have recently announced store closures and have mothballed stores that they have built but not opened. There have been calls from politicians for them to deplete their huge land banks and release this for housing. In car retailing we have witnessed not only a significant reduction in the number of dealers but also a shift in their location from the centres of cities to retail parks or retail strips where land is cheaper and customer parking a lot easier. The traditional UK high street seems under constant threat as consumers prefer the convenience of retail parks, again where parking is easier and usually free. In retail banking the demise of the high street branch has long been predicted as customers shift their transactional activity to the internet or the phone. However, according to a report from Accenture (2014), nearly 60 per cent of traditional bank products in USA and Canada are sold via the branch and 66 per cent of customers still prefer to 'talk to a person' rather than purchase online. Although the number of bank branches in the UK has fallen from 13,349 in 1997 to 9,500 by 2014, there has been major investment in the remaining branches with 2,300 branches having been refurbished between 2012 and 2014. Nationwide, the UK's largest building society, announced in March 2015 that it was to invest £300 million to upgrade its 700 branch network.

The conclusion from this fast changing environment is that that network 'reinvention' (cf. Chapter 9) holds the key for retailers to successfully take advantage of their investment in their bricks-and-mortar estate. As discussed in Chapter 10, new innovations, such as click and collect, can breathe life into the established physical network for many retailers. However, those failing to embrace change in this market are likely to face major difficulties. What these examples also demonstrate is that while retailing is a fast moving sector the importance of a sound, evidence based, understanding of changing customer and competitor behaviour remains essential.

12.2 Growth of data

We discussed in some detail in Chapter 11 how the interest in the potential of big data to assist in retail and consumer analysis has grown rapidly in recent years. However, it is fair to remark that big data has been around for longer than many care to remember. For example, banks have been collecting customer transaction data for many decades and supermarkets introduced Electronic Point of Sale (EPOS) systems in the 1980s. In the 1990s a whole new industry emerged under the heading of 'Data mining' which heavily centred around the financial service industry and was supported by equipment manufacturers such as IBM and Terradata. One of the problems then was that the hardware costs to process very large data sets were expensive and the analytical skills required to produce meaningful and actionable results were thin on the ground. This latter point remains the case today.

Supermarkets have enhanced the value of their EPOS data by linking it with loyalty card data. Only Asda among the big UK four does not now have a loyalty card. But it also worth noting that both Tesco and Sainsbury's launched their loyalty cards in the 1990s. The concept of incentivising loyalty goes back even further. In the 1960s, cigarette manufacturers issued coupons with their packets for smokers to collect and redeem for gifts. Older readers will also remember Green Shield Stamps, which certain retailers provided that could be collected in books and again redeemed for gifts. Airlines have been using loyalty cards since the 1980s where airline partners provide rewards to customers who remain loyal to their own airline and group partnerships.

What has perhaps generated most interest in the past couple of years is the potential to make use of mobile phone and social media data to better understand consumer behaviour and preferences. We began looking at this in Chapter 11 but it's probably fair to say that at the time of writing the jury is still out on how effective the analysis of these data sets are for retailers. What is required is further empirical research on the application of these types of data sets to real world retail issues.

12.3 Omni-channel retailing

Like big data, omni-channel retailing is perceived by many as a relatively recent phenomena driven by the growth of online retailing and mobile devices. In reality consumers have had choices of which channel to buy products for decades – of course the channels were different but there was still choice. Mail order catalogues from companies such as Kays and Littlewoods (merged in 2005 under the name of 'Shop Direct') were established in the nineteenth and early twentieth centuries respectively and since then have typically sold clothing and durable goods in both a spring/summer and an autumn/winter collection. Also in the 1950s and 1960s in the UK it was common for families to submit an order for groceries via an order book and these would be delivered to your home in a box on a specified day. Not quite Tesco.com but the same idea. In the same era it was not unusual for door-to-door salesmen to frequently visit homes selling all sorts of things.

However, the retail world has significantly changed since the advent of the internet and the emergence of online shopping. As we discussed in Chapter 10, it works better for some products than others. A latest best-selling novel will appear exactly the same if you buy it in a shop or online. The same does not necessarily hold true for other products where consumers may wish to express their preference over a cut of meat or a type of fish. The growth of online shopping has led to the emergence of retailers with no physical outlets, such as ASOS (clothing) and Ocado (groceries) in the UK. But we should bear in mind that the first branchless bank in the UK, First Direct was set up in 1989 before the internet had gained traction. Initially using call centres as the means for communicating with customers and ATMs as a channel for cash withdrawal it proved very popular. Of course it now has an online banking facility but it does remind us that there was a lot of innovation in retailing before online shopping took hold. Indeed in financial services, multi-channel formats have been the norm for several decades and conferences on the challenges this represents have been held since the mid-1990s.

The point we are trying to make here (and in Chapters 10 and 11) is that most consumers will continue to use a mix of channels to consume retail products and services depending on a range of factors (geography being an important one). The retailers that embrace multi-channel retailing are likely to be the most successful in the long run.

12.4 Growth of convenience

By convenience we don't just mean convenience stores, although they are a major component of this trend. We can widen the definition to include customer convenience, making shopping easier, efficient, cost and time effective. This has come as a result of an increasingly competitive retail environment in which the retailer's struggle to maintain and increase market share results in them having to do business on their customers' terms of engagement. This manifests itself in many different ways. First, many retail stores have extended their opening hours. Many UK supermarkets are open 24 hours a day, six days a week (and are likely to have extended hours on a Sunday in the future). Second, many retailers offer price promises by guaranteeing to match or beat prices of their main competitors. Thus potentially you don't have to shop around for the best deal. Similarly, comparison websites allow consumers to seek advice on best buys for different types of products (although we note that the consumer publication *Which?* has been doing this for several decades, originally in a monthly magazine). Comparison sites essentially take the hassle out of having to visit several different outlets (maybe in different locations) to find the best deal on, say, a new washing machine or car insurance etc. Third, online shopping allows consumers to purchase goods at their time of choosing, in the comfort of their own homes and delivered, in some cases, at a time of their choosing.

The final trend we point to is, indeed, the growth of convenience stores. These fall into two main categories in the UK: symbol groups and branded stores. The former, including Spar and Mace, are run by independent retailers who operate under the symbol group banner. This provides access to stronger purchasing power, marketing and merchandising skills. The branded c-stores are those operated by the larger supermarket groups, notably Tesco and Sainsbury's but also now including Waitrose and Marks and Spencer. We discussed the growth of these types of stores in Chapter 2.

12.5 Complexity of consumer behaviour

Intertwined with trends identified above are changes in consumer behaviour in relation to product purchase, brand and channel preferences, frequency of purchase and so on. This change has been brought on by a number of factors discussed throughout the book. To recap, these include an increase in the number of elderly people in the UK; more diversity in household compositions, including an increase in the number of single-person households, an increase in migration into the UK, especially from Eastern Europe, and a growth in the number of people from non-white ethnic backgrounds. Importantly, none of these changes are taking place uniformly across the country. Population growth is strongest in the south-east, concentrations of ethnic minority groups are focused on cities such as Leicester and Bradford, and Eastern European migrants are most commonly found in the east of England. Therefore geographical analyses of these trends are important, especially looking at how the changes will pan out in the future and the implications they hold in store for retailers.

Another significant change that has impacted on consumer behaviour has been the UK entering a (global) economic recession in 2008. In real terms household income in the UK has declined during the intervening years between then and now. This has forced changes in retail spending behaviour. Analysis using Acxiom data on household expenditure patterns across the UK (Thompson *et al.* 2014) suggests that there has been a marked shift from eating out to eating at home which, in a strange way, benefits supermarkets. But perhaps most interesting, the analysis in Thompson *et al.* (2014) shows a marked shift towards consumers shopping at the 'deep discounters' such as Aldi and Lidl, which is reflected in these stores' aggressive expansion programmes. While the UK big four have put most of their new superstore openings on ice, Aldi and Lidl continue to open new stores. Indeed, in March 2015 Aldi's market share of the UK grocery market overtook that of Waitrose and the Co-op to propel it into fifth position.

12.6 Continued importance of analytics

We hope we have demonstrated throughout this book that despite all the changes to the supply and demand components of the retail system, the need for geographical analysis remains strong and, arguably because of these changes, stronger than ever. In addition we believe analytics or spatial analysis will be fundamental in addressing a number of key issues in the future:

1. Impacts of demographic change on retailing. We know that the population of the UK is growing quite rapidly (the fastest growth rate in Europe). This is underpinned by increased longevity, increased in-migration from (mainly) the EU and an increase in birth rates. The implications of this change for retailers could be quite profound – e.g. the growth in older people presents opportunities to tailor offerings aimed at this sector who have different preferences than younger people.
2. Growth of convenience retailing beyond the grocery sector. This might also include the growth of 'click and collect' in a variety of retail sectors.
3. Retailing and big data. Using consumer data in far more applications than to date – perhaps also linking consumer data with health data to offer new insights into diet and consumption for example.
4. Retailing in the digital age. There has been rapid growth in the use of smart phones and tablets as a channel for product purchase. How is this likely to change in the future as mobile devices get access to the new

generation of superfast access speeds (5G), and retailers aim to tailor product recommendations and offers to individual customers?

5 The future of retail infrastructure. Many supermarkets now seem to have excess capacity in terms of their large format outlets. Many high streets have seen their vacancy rates increase over the last decade. What will the future retail landscape look like?

6 Sustainable retailing in a low carbon economy. How do retailers adapt their supply chain to reduce travel distance, minimise waste and so on? Also how proactive will they be in a new era of 'corporate social responsibility'.

These are just some of the major issues we would want to explore in the next wave of our research. We hope we might be able to share the results with readers in the next edition of this book!

References

Alexander N., Doherty A.M. (2009) *International retailing*, Oxford University Press, Oxford.

Alon, I., Qi, M., Sadkowski, R. (2001) Forecasting aggregate retail sales: a comparison of artificial neural networks and traditional methods, *Journal of Retailing and Consumer Services*, *8*, 147–156.

Amcoff, J. (2016) The geographical redistribution of retail outlets in Sweden 1998–2008, *The International Review of Retail, Distribution and Consumer Research*, *26*(3), 242–259.

Apparicio, P., Cloutier, M.S., Shearmur, R. (2007) The case of Montreal's missing food deserts: evaluation of accessibility to food supermarkets, *International Journal of Health Geographics*, *6*(1), 4.

Applebaum, W. (1966) Guidelines for a store location strategy study, *Journal of Marketing*, *30*, 42–55.

Arentze, T.A., Oppewal H., Timmermans, H.J.P. (2005) A multipurpose shopping trip model to assess retail agglomeration effects, *Journal of Marketing Research*, 109, vol. *XLII* (February), 109–115.

Ashby, D., Longley, P. (2005) Geocomputation, geodemographics and resource allocation for local policing, *Transactions in GIS*, *9*, 53–72.

Ashworth, G., Tunbridge, J.E. (1990) *The tourist-historic city*, Belhaven, London.

Aspinall, P. J. (2009) *Estimating the size and composition of the lesbian, gay and bisexual population in Britain. Research Report 37*, Equality and Human Rights Commission, London.

Bai, X., Chen, G., Tian, Q., Yin, W., Dong, J. (2009) Semi-supervised regression for evaluating convenience store location, IJCAI'09 *Proceedings of the 21st international joint conference on artificial intelligence*, Morgan Kaufmann, San Francisco, CA.

Baker, S. (2012) From the criminal crowd to the 'mediated crowd': the impact of social media on the 2011 English riots, *Safer Communities*, *11*(1), 40–49.

Ballas, D., Campbell, M., Clarke, G.P., Hanaoka, K., Nakaya, T., Waley, P. (2013) A spatial microsimulation approach to small area income estimation in Britain and Japan, *Studies in Regional Science*, *42*(1), 163–187.

Ballas, D., Clarke, G.P. (2001) Modelling the local impacts of national social policies: a spatial microsimulation approach, *Environment and Planning C, Government and Policy*, *19*, 587–606.

Ballas, D., Clarke, G.P., Dorling, D., Eyre, H., Thomas, B., Rossiter, D. (2005) SimBritain: a spatial microsimulation approach to population dynamics, *Population, Space and Place*, *11*, 13–34.

Ballas, D., Clarke, G.P., Dorling, D., Rossiter, D. (2007) Using SimBritain to model the geographical impact of National Government policies, *Geographical Analysis*, *39*(1), 44–77.

Ballas, D., Clarke, G.P., Franklin, R., Newing, A. (2017) *GIS for the social sciences: theory and application*, Taylor and Francis, Abingdon (forthcoming).

Baron, S., Harris, K., Leaver, L., Oldfield, B.M. (2001) Beyond convenience: the future for independent food and grocery retailers in the UK, *International Review of Retail, Distribution and Consumer Research*, *11*(4), 395–414.

Batty, M. (1976) *Urban modelling: algorithms, calibrations, predictions*, Cambridge University Press, Cambridge.

Beaumont, J.R. (1991) GIS and market analysis, in Maguire, D.J., Goodchild, M.F, Rhind, D.W. (eds) *Geographical information systems: principles and applications* (vol. 2), Longman, London, 139–151.

Bell, W. (1953) The social areas of the San Francisco Bay Region, *American Sociological Review*, *18*(1), 39–47.

Ben-Akiva, M.E., Lerman, S.R. (1985) *Discrete choice analysis: theory and application to predict travel demand* (vol. 9), The MIT Press, Cambridge, MA.

Ben Moussa, M. (2013) From Arab street to social movements: re-theorising collective action and the role of social media in the Arab spring, *Westminster Papers in Communication and Culture*, *9*(2), 47–70.

Benoit, D., Clarke, G.P. (1997) Assessing GIS for retail location planning, *Journal of Retailing and Consumer Services*, *4*(4), 239–258.

Bermingham, J., Baker, K., McDonald, C. (1979) ACORN – A Classification Of Residential Neighbourhoods, *Admap*, *May*, 203–209.

Bertuglia, C.S., Clarke, G.P., Wilson, A.G. (eds) (1994) *Modelling the city: performance, policy and planning*, Routledge, London.

Birkin, M., Boden, P., Williams, J. (2003) Spatial decision support systems for petrol forecourts, in Geertman, S. and Stillwell, J. (eds) *Planning support systems in practice*, Springer, Berlin, 291–304.

Birkin, M., Clarke, G.P. (2012) Enhancing spatial microsimulation using geodemographics, *Annals of Regional Science*, *49*(2), 515–532.

Birkin, M., Clarke, G.P., Clarke, M. (2002) *Retail geography and intelligent network planning*, Wiley, Chichester.

Birkin, M., Clarke, G.P., Clarke, M. (2010) Refining and operationalising entropy maximizing models for business applications, *Geographical Analysis*, *42*(4), 422–445.

Birkin, M., Clarke, G.P., Clarke, M., Culf, R. (2003) Using spatial models to solve difficult retail location problems, in Stillwell, J. and Clarke, G.P. (eds) *Applied GIS and spatial modelling*, Wiley, Chichester, 35–54.

Birkin, M., Clarke, G.P., Clarke, M., Wilson, A. (1996) *Intelligent GIS: location decisions and strategic planning*, Geoinformation International, Cambridge.

Birkin, M., Clarke, G.P., Douglas, L. (2002) Optimising retail mergers geographically, *Progress in Planning*, *58*(4), 229–318.

Birkin, M., Clarke, M. (1989) The generation of individual and household incomes at the small area level using synthesis, *Regional Studies*, *23*(6), 535–548.

Birkin, M., Clarke, M., George, F. (1995) The use of parallel computers to solve nonlinear spatial optimisation problems: an application to network planning, *Environment and Planning A*, *27*(7), 1049–1068.

Birkin, M., Foulger, F. (1992) Sales Performance and Sales Forecasting Using Spatial Interaction Modelling, School of Geography, University of Leeds, Working Paper.

Birkin, M., Harland, K., Malleson, N., Martin, D. (2013) *Microsimulation of daily movement patterns in a British city*, MODSIM, Adelaide.

Birkin, M., Harland, K., Malleson, M., Clarke, M., Cross, P. (2014) An examination of personal mobility patterns in space and time using Twitter, *International Journal of Agricultural and Environmental Information Systems*, *5*(3), 55–72.

Bloomberg News (2014) Office Depot closing 400 stores following OfficeMax merger, 6 May, www.bloomberg.com/news/2014-05-06/office-depot-will-close-400-stores-following-officemax-merger.html (accessed 20 November 2016).

Booth, Charles (undated) online archive, www.booth.lse.ac.uk/ (accessed 1 December 2016).

Brenkers, R., Verboven, F. (2006) Liberalising a distribution system: the European car market, *Journal of the European Economic Association*, *4*, 216–251.

Brown, P. (1988) A super profile based affluence ranking of OPCS urban areas, *Built Environment*, *14*, 118–128.

Brown, P. (1991) Exploring geodemographics, in Masser, I., Blakemore, M. (eds) *Handling geographical information*, Longman, London, 221–258.

Brown, P., Batey, P. (1994) The design and construction of a geodemographic targeting system: Super Profiles 1991, Working Paper No. 40, Department of Civic Design, University of Liverpool.

Buckley, M., Cowan, C., McCarthy, M. (2007) The convenience food market in Great Britain: convenience food lifestyle (CFL) segments, *Appetite*, *49*(3), 600–617.

Burns, L. (2014) Creating a Classification at the Level of the Individual, unpublished PhD thesis, School of Geography, University of Leeds.

Burt, S. (1984) Hypermarkets in France: has the Loi Royer had any effect?, *International Journal of Retail & Distribution Management*, *12*(1), 16–19.

Burt, S. (1986) The Carrefour Group – the first 25 years, *International Journal of Retailing*, *1*(3), 54–78.

Burt, S., Davies, K., Dawson, J., Sparks, L. (2008) Categorizing patterns and processes in retail grocery internationalisation, *Journal of Retailing and Consumer Services*, *15*(2), 78–92.

Burt, S., Dawson, J., Sparks, L. (2003) Failure in international retailing: research propositions, *The International Review of Retail, Distribution and Consumer Research*, *13*(4), 355–373.

Burt, S., Limmack, R. (2001) Takeovers and shareholders returns in the retail industry, *International Review of Retail, Distribution and Consumer Research*, *11*(1), 1–22.

Burt, S., Sparks, L. (1994) Structural change in grocery retailing in Great Britain: a discount reorientation?, *International Review of Retail, Distribution and Consumer Research*, *4*(2), 195–217.

Burt, S., Sparks, L. (1995) Understanding the arrival of limited line discount stores in Britain, *European Management Journal*, *13*(1), 110–119.

Business Insider (2013) Here's how some brands have subtly won over the LGBT community, www.businessinsider.com/lgbt-community-untapped-market-consumer-brands-2013-6?IR=T (accessed 20 November 2016).

CACI (1998) *Geodemographic pocketbook*, NTC Publications, Henley.

CACI (2016) *The Acorn user guide*, CACI, London.

Cairncross, F. (1997) *The death of distance*, Orion, London.

Carter, M. (2014) How Twitter may have helped Nigeria contain Ebola, *British Medical Journal*, *349*: g6496.

CEBR (2012) Data equity: unlocking the value of big data, Centre for Economics and Business Research, Report for SAS, April 2012, www.sas.com/offices/europe/uk/downloads/data-equity-cebr.pdf (accessed 20 November 2016).

Centre for Retail Research (2015) *Online Retailing: Britain, Europe, US and Canada 2015*, Centre for Retail Research, Newark, www.retailresearch.org/onlineretailing.php (accessed 20 November 2016).

China Internet Watch (2014) China online retail market grows 3 times faster than retail market, www.chinainternetwatch.com/11542/online-retail-market-expected-to-increased-by-25-yoy/#ixzz3NBW2aLgT (accessed 20 November 2016).

Choi, S.C., Dawson, J., Larke, R., Mukoyama, M. (eds) (2002) *The internationalisation of retailing in Asia*, Routledge, London.

Christopherson, S. (2007) Barriers to 'US style' lean retailing: the case of Wal-Mart's failure in Germany, *Journal of Economic Geography*, *7*(4), 451–469.

Church, R.L., Revelle, C. (1974) The maximal covering location problem, *Papers of the Regional Science Association*, *32*, 101–118.

Church, R.L. Sorensen, P. (1996) Integrating normative location-allocation models into GIS: problems and prospects with the p-median model, in Longley, P. and Batty, M. (eds) *Spatial analysis: modelling in a GIS environment*, Pearson Professional Ltd., Cambridge, 167–184.

Clarke, G.P. (ed.) (1996) *Microsimulation for urban and regional policy analysis*, Pion, London.

Clarke, G.P. (1998) Changing methods of location planning for retail companies, *GeoJournal*, *45*(4), 289–298.

Clarke, G.P., Bennison, D. (eds) (2004) Extending the food desert debate, *International Journal of Retail and Distribution Management*, *32*(2), 69–136.

Clarke, G.P., Clarke, M. (2001) From theory to practice – making spatial interaction models work in different contexts, in Clarke, G.P. and Madden, M. (eds) *Regional science in business*, Springer-Verlag, Berlin, 137–158.

Clarke, G.P., Eyre, H., Guy, C. (2002) Deriving indicators of access to food retail provision in British cities: studies of Leeds, Bradford and Cardiff, *Urban Studies*, *39*(11), 2041–2060.

Clarke, G.P., Thompson, C., Birkin, M. (2015) Exploring the geography of e-commerce in UK retailing, *Regional Studies, Regional Science*, *2*(1), 370–390.

Clarke, G.P., Wilson, A.G. (1994) A new geography of performance indicators for urban planning, in Bertuglia, C.S., Clarke, G.P. and Wilson, A.G. (eds) *Modelling the city: performance, policy and planning*, 55–82.

Clarke, I., Hallsworth, A., Jackson, P., De Kervenoael, R., Perez del Agula, R., Kirkup, K. (2006) Retail restructuring and consumer choice 1. Long-term local changes in consumer behaviour: Portsmouth, 1980–2002, *Environment and Planning A*, *38*, 25–46.

Clarke, I., Mackaness, W., Ball, B. (2003) Modelling intuition in retail site assessment (MIRSA): making sense of retail location using retailers' intuitive judgements as a support for decision-making, *The International Review of Retail, Distribution and Consumer Research*, *13*(2), 175–193.

Clarke, M. (2007) The impact of the growth in home delivery on the demand for medium and light commercial vehicles, a report for Ford of Britain, Copy available from the author, School of Geography, University of Leeds, Leeds, LS2 9JT.

Clarke, M., McDonald, A., Boden, P. (2012) Examining the relationship between debt and deprivation in the UK water industry, *Applied Spatial Analysis and Policy*, *5*, 1–22.

Clarke, R., Felson, M. (1993) *Routine activity and rational choice*, Transaction, New Brunswick, NJ.

Clarkson, R.M., Clarke-Hill, C.M., Robinson, T. (1996) UK supermarket location assessment, *International Journal of Retail & Distribution Management*, *24*(6), 22–33.

Cliff, A. (1981) *Spatial diffusion: an historical geography of epidemics in an island community*, Cambridge University Press, Cambridge.
Cliquet, G. (2000) Large format retailers: a French tradition despite reactions, *Journal of Retailing and Consumer Services*, *7*(4), 183–195.
Coe, N.M., Lee, Y.S. (2006) The strategic localization of transnational retailers: the case of Samsung–Tesco in South Korea, *Economic Geography*, *82*(1), 61–88.
Coe, N.M., Lee, Y.S. (2013) 'We've learnt how to be local': the deepening territorial embeddedness of Samsung–Tesco in South Korea, *Journal of Economic Geography*, *13*(2), 327–356.
Coe, N.M., Wrigley, N. (2009) *The globalisation of retailing*, Edward Elgar, Cheltenham.
Colla, E. (2003) International expansion and strategies of discount grocery retailers: the winning models, *International Journal of Retail & Distribution Management*, *31*(1), 55–66.
Colla, E. (2004) The outlook for European grocery retailing: competition and format development, *The International Review of Retail, Distribution and Consumer Research*, *14*(1), 47–69.
Coltman, T., Devinney, T.M., Latukefu, A., Midgley, D.F. (2001) E-business: revolution, evolution, or hype?, *California Management Review*, *44*(1), 57–86.
Competition Commission (2011) *Commentary on retail mergers*, UK Competition Commission, March 20111, Office of Fair trading, London.
Coombes, M.G., Green, A. ., Openshaw, S. (1986) An efficient algorithm to generate official statistical reporting areas: the case of the 1984 travel-to-work areas revision in Britain, *Journal of the Operational Research Society*, 943–953.
Cummins, S., Pettigrew, M., Higgins, C., Findlay, A., Sparks, L. (2005) Large-scale food retailing as an intervention for diet and health, *Journal of Epidemiology and Community Health*, *59*, 1035–1040.
Das, P., Chaudhury, S. (2007) Prediction of retail sales of footwear using feedforward and recurrent neural networks, *Neural Computing & Applications*, *16*(4–5), 491–502.
Data Economy (2014) Understanding big data: the seven V's, *Data Economy*, 22 May.
Dawson, J., Kirby, D. (1979) *Small scale retailing in the UK*, Saxon House, Farnborough.
Dawson, J., Larke, R., Mukoyama, M. (eds) (2006) *Strategic issues in international retailing*, Routledge, London.
Debenham, J., Clarke, G.P., Stillwell, J.C.H. (2003a) Extending geodemographics: a new regional prototype, *Environment and Planning A*, *35*, 1025–1050.
Debenham, J., Stillwell, J.C.H., Clarke, G.P. (2003b) The estimation of self-containment and catchment size indicators for use in small-area classification, *International Journal of Population Geography*, *9*, 253–271.
De Kervenoael, R., Hallsworth, A., Clarke, I. (2006) Macro-level change and micro level effects: a twenty-year perspective on changing grocery shopping behaviour in Britain, *Journal of Retailing and Consumer Services*, *13*, 381–392.
de Mello-Sampayo, F. (2009) Competing-destinations gravity model: an application to the geographic distribution of FDI, *Applied Economics*, *41*(17), 2237–2253.
Deloitte (2013) *Tourism: jobs and growth*, Deloitte, London.
Department of the Environment (1987) *Handling geographical information: report to the Secretary of State for the Environment of the Committee of Enquiry into the handling of geographical information (the Chorley report)*, HMSO, London.
Department of Transport (2014) Vehicle licensing statistics 2103, statistical release, 10 April 2014, www.gov.uk/government/uploads/system/uploads/attachment_data/file/302409/vls-2013.pdf (accessed 20 November 2016).

Dudding, V., Ryan, C. (2000) The impacts of tourism on a rural retail sector: a New Zealand case study, *Tourism Economics*, *6*(4), 301–319.

Duggal, N. (2008) *Use of GIS in retail location analysis: Burger King & McDonald's in Portage & Summit Counties*, VDM, Ohio.

Duhigg, C. (2012) How companies learn your secrets, *New York Times*, 16 February.

Eckert, J., Shetty, S. (2011). Food systems, planning and quantifying access: using GIS to plan for food retail. *Applied Geography*, *31*(4), 1216–1223.

Edwards, J. (1975) Social indicators, urban deprivation and positive discrimination, *Journal of Social Policy*, *4*(3), 275–287.

Enders, A., Jelassi, T. (2000) The converging business models of internet and bricks-and-mortar retailers, *European Management Journal*, *18*(5), 542–550.

Epstein, J. (2009) Modelling to contain pandemics, *Nature*, *460*(7256), 687.

ESRC (2012) *Improving access for research and policy: report from the Administrative Data Task Force, December 2012*, Economic and Social Research Council, Swindon.

Eurostat (2015) Air passenger transport – monthly statistics, Eurostat, Brussels, ec.europa.eu/eurostat/statistics-explained/index.php/Air_passenger_transport_-_monthly_statistics (accessed 20 November 2016).

Eyre, H.A. (1999) *Measuring the performance of spatial interaction models in practice*, unpublished PhD thesis, University of Leeds.

Fagan, M. (1999) *Northern exposure*, *Sunday Telegraph*, 4 April 1999.

Farag, S., Weltevreden, J, van Rietbergen, T., Dijst, M. van Ort, F. (2006) e-shopping in the Netherlands: does geography matter?, *Environment and Planning B*, *33*, 59–74.

Farhan, B., Murray, A.T. (2006) Distance decay and coverage in facility location planning, *The Annals of Regional Science*, *40*(2), 279–295.

Fenwick, I. (1978) *Techniques in store location research: a review and applications*, Retail and Planning Associates, Corbridge.

Ferguson, N., Cummings, A., Cauchemez, S., Fraser, C., Riley, S., Meeyai, A., Iamsirithaworn, S., Burke, D. (2005) Strategies for containing an emerging influenza pandemic in Southeast Asia, *Nature*, *437*, 209–214.

Fernie, J. (2009) *Logistics and retail management: emerging issues and new challenges in the retail supply chain*, Kogan Page, London.

Fernie, J., Arnold, S. J. (2002) Wal-Mart in Europe: prospects for Germany, the UK and France, *International Journal of Retail & Distribution Management*, *30*(2), 92–102.

Fernie, J., Moore, C., Lawrie, A., Hallsworth, A. (1997) The internationalization of the high fashion brand: the case of central London, *Journal of Product & Brand Management*, *6*(3), 151–162.

Fernie, J., Sparks, L. (2009) Retail logistics: changes and challenges, *Logistics and Retail Management*, 3–37.

Flowerdew, R., Green, M., Kehris, E. (1991) Using areal interpolation methods in geographic information systems, *Papers in Regional Science*, *70*(3), 303–315.

Fotheringham, A.S. (1983) A new set of spatial interaction models: the theory of competing destinations, *Environment and Planning A*, *15*(1), 15–36.

Fotheringham, A.S. (1986) Modelling hierarchical destination choice, *Environment and Planning A*, *18*, 401–418.

Fotheringham, A.S. (1988) Note – consumer store choice and choice set definition, *Marketing Science*, *7*(3), 299–310.

Fotheringham, A.S. (2013) Localised spatial interaction modelling, paper presented at IGU Applied Geography Workshop, Leeds, May 2013, available at www.youtube.com/watch?v=4GDb9PyzxXg (accessed 20 November 2016).
Fotheringham, A.S., Nakaya, T., Yano, K., Openshaw, S., Ishikawa, Y. (2001) Hierarchical destination choice and spatial interaction modelling: a simulation experiment, *Environment and Planning A*, *33*(5), 901–920.
Fotheringham, A.S., Trew, R. (1993) Chain image and store-choice modeling: the effects of income and race, *Environment and Planning A*, *25*(2), 179–196.
Freathy, P., O'Connell, F. (1998) *European airport retailing*, Macmillan, London.
Gao, Y.F., Liang, Y.S., Tang, F., Ou, Z.W., Peng, Y.J., Liang, J. (2010) Research on a neural-network-based forecasting algorithm for retail industry, in *Intelligent control and automation (WCICA), 2010 8th World Congress*). IEEE. 5201–5205.
Gen, M., Cheng, R. (2000) *Genetic algorithms and engineering optimization*, John Wiley & Sons, London.
George, F., Radcliffe, N., Smith, M., Birkin, M., Clarke, M. (1997) Spatial interaction model optimisation on parallel computers, *Concurrency: Practice and Experience*, *9*(8), 753–780.
Ghosh, A., Craig, C.S. (1984) A location-allocation model for facility planning in a competitive environment, *Geographical Analysis*, *16*, 39–56.
Ghosh, A., Harche, F. (1993) Location-allocation models in the private sector: progress, problems and prospects, *Computers & Operations Research*, *1*(1), 81–106.
Ghosh, A., McLafferty, S.L. (1987) *Location strategies for retail and service firms*, Lexington Books, Massachusetts.
Ginsberg, J., Mohebbi, M., Patel, R., Brammer, L., Smolinski, M., Brilliant, L. (2009) Detecting influenza epidemics using search engine query data, *Nature*, *457*(7232), 1012–1014.
Glendenning, M., Page, D. (1999) *Clone city*, Polygon, Edinburgh.
Gofton, L. (1995) Convenience and the moral status of consumer practices, in Marshall, D.W. (ed.) *Food choice and the consumer*, Chapman and Hall, New York.
Goldman, A. (2001) The transfer of retail formats into developing economies: the example of China, *Journal of Retailing*, *77*(2), 221–242.
González-Benito, B., González-Benito, J. (2005) The role of geodemographic segmentation in retail location strategy, *International Journal of Market Research*, *47*(3), 295–316.
Gonzáles-Benito, O., Gonzáles-Benito, J. (2004) Geographic price discrimination as a retail strategy: role of the geodemographic market profile, *International Journal of Market Research*, *46*(4), 443–461.
Goodchild, M. (1984) ILACS: a location-allocation model for retail site selection, *Journal of Retailing*, *60*(1), 84–100.
Goss, J. (1995) 'We know who you are and we know where you live': the instrumental rationality of geodemographic systems, *Economic Geography*, 171–198.
Graff, T.O. (1998) The locations of Wal-Mart and Kmart supercenters: contrasting corporate strategies, *The Professional Geographer*, *50*(1), 46–57.
Graff, T.O., Ashton, D. (1994) Spatial diffusion of Wal-Mart: contagious and reverse hierarchical elements, *The Professional Geographer*, *46*(1), 19–29.
Grewal, D., Levy, M., Lehmann, D. (2004) Retail branding and customer loyalty: an overview, *Journal of Retailing*, *80*, 249–252.
Guardian (undated) University Guide, *The Guardian* online, www.theguardian.com/education/universityguide (accessed 20 November 2016).

Guardian (2013) Tesco's empire: expansion checked in UK and beyond, *Guardian*, 18 April 2013, www.theguardian.com/business/2013/apr/18/tesco-empire-uk-and-beyond (accessed 20 November 2016).

Gulati, R., Garino, J. (2000) Get the right mix of bricks and clicks, *Harvard Business Review*, *78*, 107–114.

Guy, C.M. (1996) Grocery store saturation in the UK – the continuing debate, *International Journal of Retail & Distribution Management*, *24*(6), 3–10.

Guy, C.M. (1998) Controlling new retail spaces: the impress of planning policies in Western Europe, *Urban Studies*, *35*(5–6), 953–979.

Guy, C.M. (2006) *Planning for retail development: a critical view of the British experience*, Taylor & Francis, London.

Guy, C.M. (2008) Assessing the Competition, *Town and Country Planning*, *77*, 14–16.

Guy, C.M., Clarke, G.P., Eyre, H. (2004) Food retail change and the growth of food deserts: a case study of Cardiff, *International Journal of Retail & Distribution Management*, *32*(2), 72–88.

Hall, S.M. (2011) High street adaptations: ethnicity, independent retail practices and localism in London's urban margins, *Environment and Planning A*, *43*, 2571–2588.

Hallsworth, A., De Kervenoael, R., Elms, J., Canning, C. (2010) The food superstore revolution: changing times, changing research agendas in the UK, *The International Review of Retail, Distribution and Consumer Research*, *20*(1), 135–146.

Hansen, W.G. (1959) How accessibility shapes land use, *Journal of the American Institute of Planners*, *25*(2), 73–76.

Harper, G., Williamson, I., Clarke, G.P., See, L. (2002) *Family origins: developing groups of crime and disorder reduction partnerships and police basic command units for comparative purposes*, Home Office, London.

Harris, R., Sleight, P., Webber, R. (2005) *Geodemographics, GIS and neighbourhood targeting*, John Wiley, London.

Henriques-Marques, S.H., Trindade, G., Santos, M. (2016) The importance of atmospherics in the choice of hypermarkets and supermarkets, *The International Review of Retail, Distribution and Consumer Research*, *26*(1), 17–34.

Heppenstall, A., Evans, A., Birkin, M. (2006) Using hybrid agent-based systems to model spatially-influenced retail markets, *Journal of Artificial Societies and Social Simulation*, *9*(3), 2, www.jasss.soc.surrey.ac.uk/9/3/2.html (accessed 20 November 2016).

Hernandez, T. (2007) Enhancing retail location decision support: the development and application of geovisualization, *Journal of Retailing and Consumer Services*, *14*(4), 249–258.

Hernandez, T., Bennison, D. (2000) The art and science of retail location decisions, *International Journal of Retail & Distribution Management*, *28*(8), 357–367.

Heywood, I., Cornelius, S., Carver, S. (2011) An introduction to GIS, *Pearson Education*, Harlow.

Hill, K. (2012) How Target figured out a teen girl was pregnant before her father did, *Forbes Magazine*, 16 February.

Hollingsworth, A. (2004) Increasing retail concentration: evidence from the UK food retail sector, *British Food Journal*, *106*(8), 629–638.

Hood, N., Clarke, G.P., Clarke, M. (2016) Segmenting the growing UK convenience store market for retail location planning, *The International Review of Retail, Distribution and Consumer Research*, *26*(2), 113–136.

Hood, N., Clarke, G.P., Clarke, M. (2017) Testing alternative revenue forecasting methods in the food convenience market, School of Geography, University of Leeds (forthcoming working paper).

Hotelling, H. (1929) Stability in Competition, *Economic Journal*, *39*, 41–57.

House of Commons (2014) Social media firms told to simplify terms and conditions, Report on Responsible Use of Data, Science and Technology Committee, 19 November.

Howe, S. (ed.) (2003) *Retailing in the European Union*, Routledge, London.

Hu, P., Pooler, J. (2002) An empirical test of the competing destinations model, *Journal of Geographical Systems*, *4*(3), 301–323.

Hubbard, N. (1999) *Acquisition*, Macmillan, Basingstoke.

Huff, D.L. (1963) A probabilistic analysis of shopping center trade areas, *Land Economics*, *39*(1), 81–90.

Huff, D.L. (1964) Defining and estimating a trading area, *Journal of Marketing*, 34–38.

Hughes, R., Birkin, M., Clarke, G.P. (2017) Exploring alternative retail distribution systems to minimise food miles for local products, School of Geography, University of Leeds (forthcoming working paper).

Hughes, R., Hallsworth, A.G., Clarke, G.P. (2009) Testing the effectiveness of the proposed UK 'competition test', *The Service Industries Journal*, *29*(5), 569–590.

Humby, C., Hunt, T. (2003) *Scoring points: how Tesco is winning customer loyalty*, Kogan Page, London.

Humby, C., Hunt, T., Phillips, T. (2008) *Scoring points: how Tesco is winning customer loyalty* (2nd edition), Kogan Page, London.

IGD (2015) *Convenience retailing factsheet*, IGD, London, www.igd.com/Research/Retail/Convenience/3369/Convenience-Retailing-Market-Overview/ (accessed 20 November 2016).

Ireland, P. (1994) GIS: another sword for St. Michael, *Mapping Awareness*, *April*, 26–29.

Ishikawa, Y. (1987) An empirical study of the competing destinations model using Japanese interaction data, *Environment and Planning A*, *19*(10), 1359–1373.

Jackson, P., Perez dez Aguila, R., Clarke, I., Hallworth, A., De Kervenoael, R., Kirkup, M. (2006) Retail restructuring and consumer choice 2. Understanding consumer choice at the household level, *Environment and Planning A*, *38*, 47–67.

Jarman, B. (1983) Identification of underprivileged areas, *British Medical Journal*, *286*, 1705–1708.

Javelin (2015) *How many supermarkets will we really need? The future of the UK grocery market*, Javelin Group, London.

Johnston, R., Sidaway, J. (2004) *Geography and geographers: Anglo-American geography since 1945* (6th edition), Edward Arnold, London.

Kates, S.M. (2000) Out of the closet and out on the street! Gay men and their brand relationships, *Psychology & Marketing*, *17*(6), 493–513.

Kates, S.M. (2013) *Twenty million new customers! Understanding gay men's consumer behaviour,* Routledge, London.

Khawaldah, H., Birkin, M., Clarke, G.P. (2012) A critical review of retail impact assessment techniques: the case study of Silverburn in Scotland, *Town Planning Review*, *83*(2), 233–260.

Kirby-Hawkins, Birkin. M., Clarke. G.P. (2017) Exploring the geography of online grocery shoppers using retail company data, working paper, School of Geography, University of Leeds (forthcoming).

Knudsen, D.C., Fotheringham, A.S. (1986) Matrix comparison, goodness-of-fit, and spatial interaction modelling, *International Regional Science Review*, *10*(2), 127–147.

Konak, A., Coit, D.W., Smith, A.E. (2006) Multi-objective optimization using genetic algorithms: a tutorial, *Reliability Engineering & System Safety*, *91*(9), 992–1007.

Kordi, M. (2013) *Geographically weighted spatial interaction (GWSI)* (Doctoral dissertation, University of St Andrews).

Laing, R.D., Clarke, I., Mackaness, W., Ball, B., Horita, M. (2003) The devil is in the detail: visualising analogical thought in retail location decision-making, *Environment and Planning B: Planning and Design*, *30*, 15–36.

Lakshmanan, J.R., Hansen, W.G. (1965) A retail market potential model, *Journal of the American Institute of Planners*, *31*(2), 134–143.

Lambiri, D., Faggian, A., Wrigley, N. (2016) Linked-trip effects of 'town-centre-first' era foodstore development: an assessment using difference-in-differences. *Environment and Planning B: Planning and Design, 0265813515624684, first published on 28 January 2016.*

Laney, D. (2001) 3D data management: controlling data volume, velocity and variety. META Group Research Note 6.

Langston, P., Clarke, G.P., Clarke, D.B. (1997) Retail saturation, retail location, and retail competition: an analysis of British grocery retailing, *Environment and Planning A*, *29*, 77–104.

Larsen, K., Gilliland, J. (2008) Mapping the evolution of food deserts in a Canadian City, *International Journal of Health Geographics*, *7*(16) doi:10.1186/1476-072X-7–16.

Laulajainen, R. (1988) The spatial dimensions of an acquisition, *Economic Geography*, *64*, 170–187.

Leyland, A., Hegarty, R., Philips, B. (2012) *Morrisons 'looking at' Costcutter acquisition. The Grocer Online*, 16 June 2012, www.thegrocer.co.uk/companies/supermarkets/morrisons-looking-at-costcutter-acquisition/229926.article (accessed 20 November 2016).

Li, L., Revesz, P. (2004) Interpolation methods for spatio-temporal geographic data, *Computers, Environment and Urban Systems*, *28*(3), 201–227.

Lloyd, C. (2016) Are spatial inequalities growing? The scale of population concentrations in England and Wales, *Environment and Planning A*, *48*, 1318–1336.

Longley, P. (2005) Geographical information systems: a renaissance of geodemographics for public service delivery, *Progress in Human Geography*, *29*, 57–63.

Longley, P., Adnan, M., Lansley, G. (2015) The geotemporal demographics of Twitter usage, *Environment and Planning A*, *47*(2), 465–484.

Longley, P., Goodchild, M.F., Maguire, D., Rhind, D.W. (2005) *Geographical information systems: principles, techniques, applications and management*, Wiley, Chichester.

Longley, P., Webber, R., Li, C. (2008) The UK geography of the e-society: a national classification, *Environment and Planning A*, *40*(2) 362–382.

Lord, J.D. (2000) Retail saturation: inevitable or irrelevant? *Urban Geography*, *21*(4), 342–360.

Lord, J.D., Lynds, C.D. (1981) The use of regression models in store location research: a review and case study, *Akron Business and Economic Review*, *12*(2), 13–19.

Lovelace, R., Malleson, N., Cross, P., Birkin, M. (2014) From 'Big Noise' to 'Big Data': a case study of cross-validation between 3 large geographical datasets on visitor flows between regional urban centres, paper presented to North American Regional Science Council, Washington, DC, 14 November.

Lovelace, R., Malleson, N., Harland, K., Birkin, M. (2014) Can social media data be useful in spatial modelling? a case study of 'museum tweets' and visitor flows, paper presented to GIS Research UK, Glasgow.

McCarthy, P. (1980) A study of the importance of generalised attributes in shopping choice behaviour, *Environment and Planning*, *12*, 1269–1286.

Macintyre, S., Macdonald, L., Ellaway, A. (2008) Do poorer people have poorer access to local resources and facilities? The distribution of local resources by area deprivation in Glasgow, Scotland, *Social Science & Medicine*, *67*(6), 900–914.

Malleson, N., Birkin, M. (2013) Generating Individual Behavioural Routines from Massive Social Data for the Simulation of Urban Dynamics, in Gilbert, T., Kirkilionis, M., Nicolis, G. (eds) *Proceedings of the European Conference on Complex Systems 2012*, Springer International Publishing, 849–855.

Malleson, N., Birkin, M., Harland, K. (2014) New insights into individual activity spaces using crowd-sourced big data, *Proceedings of IEEE BigDataScience 2014*, Berkeley, CA.

Maraschin, C., Krafta, R. (2013) Growth dynamic of retail locations: a methodological approach using a logistic model, *Cybergeo: European Journal of Geography*, www.cybergeo.revues.org/26002 (accessed 20 November 2016).

Masser, I. (1979) A note on the treatment of flows across system boundaries in spatial interaction models, *Environment and Planning A*, *11*, 447–454.

Mattman, C. (2013) Computing: a vision for data science, *Nature*, *493*, 473–475.

Mayer-Schonberger, V., Cukier, K. (2013) *Big Data: A revolution that will transform how we live, work and think*, John Murray, London.

Meksangsouy, P. (2012) The impact of internationalisation on retailing in Bangkok and its implications for retail location models, PhD thesis, School of Geography, University of Leeds.

Mendes, A., Themido, I. (2004) Multi-outlet retail site assessment, *International Transactions in Operations Research*, *11*, 1–18.

Meneely, L., Burns, A., Strugnell, C. (2009) Age associated changes in older consumers' retail behaviour, *International Journal of Retail & Distribution Management*, *37(*12), 1041–1056.

Mintel (2003) *Attitudes towards buying local produce, market intelligence*, January 2003, Mintel International Ltd, London.

Mintel (2013) *UK vs US Online Shopping – June 2000*. Mintel International Ltd: London, www.marketresearch.com/Mintel-Snapshots-v2587/Commerce-UK-snapshot-7106170/ (accessed 20 November 2016).

MobileShopTalk (2015) Mobile moves in on grocery shopping, Media Post, www.mediapost.com/publications/article/242215/mobile-moves-in-on-grocery-shopping.html (accessed 1 December 2016).

Mokhtarian, P. (2004) A conceptual analysis of the transportation impacts of b2c e-commerce, *Transportation*, *31*, 257–284.

Moore, K., Ruddle, K. (2000) Organising for e-Business, in Wilcocks, L., Sauer, C. (eds) *Moving to e-business: the ultimate practical guide to effective e-business*, Random House, London.

Morrill, R. (1968) Waves of spatial diffusion, *Journal of Regional Science*, *8*(1), 1–18.

Morrisons Online (2011) Morrisons announces opening of first convenience store, www.morrisons.co.uk/corporate/Media-centre/Corporate-news/Morrisons-announces-opening-of-first-convenience-store/ (accessed 20 November 2016).

Mortimer, G., Fazal e Hasan, S., Andrews, L., Martin, J. (2016) Online grocery shopping: the impact of shopping frequency on perceived risk, *The International Review of Retail, Distribution and Consumer Research*, 1–22.

Moschis, G.P., Lee, E., Mathur, A. (1997) Targeting the mature market: opportunities and challenges, *Journal of Consumer Marketing*, *14*(4), 282–293.

Moser, C., Scott, W. (1961) *British towns: a statistical study of their social and economic differences*, Oliver & Boyd, Edinburgh.

Moses, L. (1958) Location and the theory of production, *Quarterly Journal of Economics*, *73*, 259–272.

Mulligan, G.F. (1983) Consumer demand and multipurpose shopping behaviour, *Geographical Analysis*, *15*(1), 76–81.

Murray, A.T. (2010) Advances in location modeling: GIS linkages and contributions, *Journal of Geographical Systems*, *12*(3), 335–354.

Myers, W. (1993) *Food Retail Perception and Reality*, Henderson Crosthwaite Institutional Brokers Ltd, 32 St Mary Hill, London EC3R 1AA.

Nakaya, T. (2001) Local spatial interaction modelling based on the geographically weighted regression approach, *GeoJournal*, *53*(4), 347–358.

Nakaya, T., Fotheringham, A.S., Hanaoka, K., Clarke, G.P., Ballas, D., Yano, K. (2007) Combining microsimulation and spatial interaction models for retail location analysis, *Journal of Geographical Systems*, *4*, 345–369.

NCRM (undated) National Centre for Research Methods website available at www.ncrm.ac.uk (accessed 20 November 2016).

NEF (2005) *Clone town Britain, Report of the New Economics Foundation*, London.

NEF (2010) *Clone town Britain 2010, Report of the New Economics Foundation*, London.

Neville, S. (2013) Morrisons adds HMV stores to its basket. *Guardian Online*, Tuesday 26 February 2013, www.guardian.co.uk/business/2013/feb/26/morrisons-buys-hmv-local-stores (accessed 20 November 2016).

Newing, A. (2013) Incorporating seasonal visitor demand in retail location modelling, unpublished PhD thesis, School of Geography, University of Leeds.

Newing, A., Clarke, G.P., Clarke. M. (2013a) Identifying seasonal variations in store-level grocery demand, *International Journal of Retail & Distribution Management*, *41*(6), 477–492.

Newing, A., Clarke, G.P., Clarke, M. (2013b) Visitor expenditure estimation for grocery store location planning: a case study of Cornwall, *International Review of Retail Distribution and Consumer Research*, *23*(3), 221–244.

Newing, A., Clarke, G.P., Clarke, M. (2014) Exploring small area demand for grocery retailers in tourist areas, *Tourism Economics* *20*(2), 407–427.

Newing, A., Clarke, G.P., Clarke, M. (2015) Developing and applying a disaggregated retail location model with extended retail demand estimations, *Geographical Analysis* *47*, 219–239.

Nicholls, A., Watson, A. (2005) Implementing e-value strategies in UK retailing, *International Journal of Retail and Distribution Management*, *33*(6), 426–443.

Oakenfull, G.K., Greenlee, T.B. (2005) Queer eye for a gay guy: using market-specific symbols in advertising to attract gay consumers without alienating the mainstream, *Psychology & Marketing*, *22*(5), 421–439.

O'Kelly, M.E. (1981) A model of the demand for retail facilities, incorporating multi-stop, multipurpose trips, *Geographical Analysis*, *13*(2), 134–148.

Oliver, M.A., Webster, R. (1990) Kriging: a method of interpolation for geographical information systems, *International Journal of Geographical Information System, 4*(3), 313–332.

ONS (2011) National Population Projections, 2010-based Statistical Bulletin, www.ons.gov.uk/ons/rel/npp/national-population-projections/2010-based-projections/stb-2010-based-npp-principal-and-key-variants.html (accessed 20 November 2016).

ONS (2012a) 2011 Census: Population Estimates for the United Kingdom, 27 March 2011, Statistical Bulletin, Office of National Statistics, London, www.ons.gov.uk/ons/dcp171778_292378.pdf (accessed 20 November 2016).

ONS (2012b) 2010-based subnational population projections for England, Office of National Statistics, London, www.ons.gov.uk/ons/rel/snpp/sub-national-population-projections/2010-based-projections/stb-2010-based-snpp.html (accessed 20 November 2016).

ONS (2012c) International Migrants in England and Wales 2011 Office of National Statistics, London, www.ons.gov.uk/ons/rel/census/2011-census/key-statistics-for-local-authorities-in-england-and-wales/rpt-international-migrants.html (accessed 20 November 2016).

ONS (2012d) 2011 Census – Population and Household Estimates for England and Wales, March 2011, Statistical Bulletin, www.ons.gov.uk/ons/dcp171778_270487.pdf (accessed 20 November 2016).

ONS (2013) National Population Projections, 2012-based Statistical Bulletin, Office of National Statistics, London, www.ons.gov.uk/ons/rel/npp/national-population-projections/2012-based-projections/stb-2012-based-npp-principal-and-key-variants.html (accessed 20 November 2016).

Openshaw, S. (1984) Ecological fallacies and the analysis of areal census data, *Environment and Planning A*, *16*(1), 17–31.

Oppewal, H., Timmermans, H.J.P., Louviere, J.J. (1997) Modelling the effects of shopping centre size and store variety on consumer choice behaviour, *Environment and Planning A*, *29*(6), 1073–1090.

Oppewal, H., Timmermans, H. (2001) Discrete choice modelling: basic principles and application to parking policy assessment, in Clarke, G.P. and Madden, M. (eds) *Regional Science in Business*, Springer, Berlin, 97–114.

Orford, S., Dorling, D., Mitchell, R., Shaw, M., Smith, G.D. (2002) Life and death of the people of London: a historical GIS of Charles Booth's inquiry, *Health & Place*, *8*, 25–35.

ORR (2016) Office of Rail and Road, UK, *Annual Statistics*, http://orr.gov.uk/statistics/published-stats/station-usage-estimates (accessed 20 November 2016).

Ottensmann, J.R. (1997) Partially constrained gravity models for predicting spatial interactions with elastic demand, *Environment and Planning A*, *29*(6), 975–988.

Pacione, M. (1974) Measures of the attraction factor; a possible alternative, *Area*, *6*, 279–282.

Pacione, M. (2005) *Urban geography: a global perspective*, Routledge, London.

Palmer, M. (2005) Retail multinational learning: a case study of Tesco, *International Journal of Retail & Distribution Management*, *33*(1), 23–48.

Palmer, M., O'Kane, P. (2007) Strategy as practice: interactive governance spaces and the corporate strategies of retail transnationals, *Journal of Economic Geography*, *7*(4), 515–535.

Papaoikonomou, E., Ryan, G., Gineis, M. (2011) Towards a holistic approach of the attitude behaviour gap in ethical consumer behaviours: empirical evidence from Spain, *International Advances in Economic Research*, *17*(1), 77–88.

Pappas, N. (2016) Marketing strategies, perceived risks, and consumer trust in online buying behaviour. *Journal of Retailing and Consumer Services*, *29*, 92–103.

Park, R., Burgess, E. (1925) *The City*, University of Chicago Press, Chicago, IL.

Pearson, J.K. (2007) A comparative business site-location feasibility analysis using geographic information systems and the gravity model, www.gis.smumn.edu/grad-projects/pearsonj.pdf (accessed 20 November 2016).

Pellegrini, P.A., Fotheringham, A.S., Lin, G. (1997) An empirical evaluation of parameter sensitivity to choice set definition in shopping destination choice models, *Papers in Regional Science*, *76*(2), 257–284.

Peñaloza, L. (1996) We're here, we're queer, and we're going shopping! A critical perspective on the accommodation of gays and lesbians in the US marketplace, *Journal of Homosexuality*, *31*(1–2), 9–41.

Pickles, J. (ed.) (1995) *Ground truth: the social implications of geographic information systems*, Guilford Press, Guildford.

Pike, R., Neale, C. (2003) *Corporate finance and investment: decisions and strategies* (4th edition), Pitman FT.

Poole, R., Clarke, G.P., Clarke, D. (2002a) Grocery retailers and regional monopolies, *Regional Studies*, *36*(6), 643–659.

Poole, R., Clarke, G.P., Clarke, D.B. (2002b) Growth, concentration and regulation in European food retailing, *European Urban and Regional Studies*, *9*(2), 167–186.

Retail Times (2010) Spar UK clinches £150m holiday park contract, *Retail Times*, 12 June.

Reynolds, J. (1997) The internet as a strategic resource: evidence from the European Retail Sector, in Willcocks, L., Leslie, P., Feeny, D., Islei, G.L. (eds) *Managing Information Technology as a Strategic Resource*, McGraw-Hill, Maidenhead.

Reynolds, J., Wood, S. (2010) Location decision making in retail firms: evolution and challenge, *International Journal of Retail & Distribution Management*, *38*(11/12), 828–845.

Roy, J.R., Thill, J.C. (2004) *Spatial interaction modelling*, Springer, Berlin.

RTPI (2004) Planning for an ageing population, RTPI, London, www.rtpi.org.uk/media/6341/08-June-2007-Ageing-Population-Report-Final-edit.pdf (accessed 20 November 2016).

Savage, M., Burrows, R. (2007) The coming crisis of empirical sociology, *Sociology*, *41*(5), 885–899.

Schofield, K., Schmidt, R.Ä. (2005) Fashion and clothing: the construction and communication of gay identities, *International Journal of Retail & Distribution Management*, *33*(4), 310–323.

Shaw, H. (2006) Food deserts: towards the development of a classification, *Geografiska Annaler B*, *88*(2), 231–247.

Shevky, E., Bell, W. (1955) *Social area analysis: theory, illustrative application and computational procedures*, Stanford University Press, Stanford, CA.

Shevky, E., Williams, M. (1949) *The social areas of Los Angeles, analysis and typology*, University of California Press, Berkley, CA.

Shields, M., Kures, M. (2007) Black out of the blue light: an analysis of Kmart store closing decisions, *Journal of Retailing and Consumer Services*, *14*(4), 259–268.

Siegfried, T. (2013) Why big data is bad for science, *Science News*, 26 November.

Silcock, E., Clarke, G.P., Clarke, D.B., Wrigley, N. (1999) Grocery provision in the USA: room for expansion?, *International Journal of Retail & Distribution Management*, *27*(1), 8–21.

Silver, N. (2012) *The signal and the noise*, Allen Lane, London.

Simkin, L. (1989) SLAM: store location assessment model – theory and practice, *Omega*, *7*, 53–58.

Simkin, L. (1990) Evaluating a store location, *International Journal of Retail and Distribution Management*, *18*, 33–38.

Singleton, A. (2010) The geodemographics of educational progression and their implications for widening participation in higher education, *Environment and Planning A*, *42*, 2560–2580.

Smithers, R. (2012) Supermarkets targeting pubs to exploit legal loophole, says Camra. *Guardian Online*, 19 November, www.guardian.co.uk/business/2012/nov/19/supermarkets-target-pubs-loophole-camra (accessed 20 November 2016).

Smoyer-Tomic, K.E., Spence, J.C., Amrhein, C. (2006) Food deserts in the prairies? Supermarket accessibility and neighborhood need in Edmonton, Canada, *The Professional Geographer*, *58*(3), 307–326.

Song, Y., Sohn, J. (2007) Valuing spatial accessibility to retailing: a case study of the single family housing market in Hillsboro, Oregon, *Journal of Retailing and Consumer Services*, *14*(4), 279–288.

Sparks, L. (1996) Space wars: Wm Low and the 'auld enemy', *Environment and Planning A*, *28*(8), 1465–1484.

Sparks, L., Burt, S. (2003) e-Commerce and the retail process: a review, *Journal of Retailing and Consumer Services*, *10*, 275–286.

Spencer, A.H. (1978) Deriving measures of attractiveness for shopping centres, *Regional Studies*, *12*, 713–726.

Spatial insights (undated) *Market Potential GIS Case Study: Market Potential Analysis using ReCAP Retail Location Data: Check Cashing and Pawn Broker*s, www.spatialinsights.com/company/papers/ (accessed 20 November 2016).

Starcount (undated) website available at www.starcount.com (accessed 20 November 2016).

Stern, N. (1999) The impact of the internet on retailing, *International Trends in Retailing*, *December*, 71–87.

Stillwell, J. (2010) Ethnic population concentration and net migration in London, *Environment and Planning A*, *42*(6), 1439–1456.

Sudarsanam, P.S. (1995) *Mergers and acquisitions*, Prentice Hall, London.

Taaffe, E., Morrill, R., Gould, P. (1963) Transport expansion in underdeveloped countries: a comparative analysis, *Geographical Review*, *53*, 4, 503–529.

Taylor, S.L., Cosenza, R.M. (2002) Profiling later aged female teens: mall shopping behavior and clothing choice, *Journal of Consumer Marketing*, *19*(5), 393–408.

Tchaban, T., Griffin, J., Taylor, M. (1998) A comparison between single and combined backpropagation neural networks in the prediction of turnover, *Engineering Applications of Artificial Intelligence*, *11*, 41–47.

TechCityNews (2014) Imperial and KPMG back £20m big data partnership, *TechCity News*, 16 July.

Telegraph (2010) Twice as many Tescos open 24/7 as police stations, *Daily Telegraph*, 21 August.

Telegraph (2013) Designing cars that women want to buy, *Daily Telegraph*, 30 May.

Telegraph (2014) Hospital records of all NHS patients sold to insurers, *Daily Telegraph*, 23 February.

Telegraph (2015) Tesco closures: is your local store getting the chop? *Daily Telegraph, 28 January 2015.*

TFL (2016) Transport for London website available at www.tfl.gov.uk (accessed 1 December 2016).

The Grocer (2010) Which way forward for the stalling discount chains? *The Grocer*, 23 October, p. 14.

This is Money.co.uk (2014) Hundreds of Carphone Warehouse and Dixons stores could shut if merger goes ahead, www.thisismoney.co.uk/money/markets/article-2567034/Dixons-3-5bn-merger-talks-Carphone-Warehouse.html (accessed 20 November 2016)

Thomas, C.J., Bromley, R.D.F., Tallon, R. (2004) Retail parks revisited: a growing competitive threat to traditional shopping centres?, *Environment and Planning A*, *36*, 647–666.

Thomas, M., Stillwell, J., Gould, M. (2013) Modelling residential mobility behaviour using a commercial data set: an analysis of mover/stayer characteristics across the life-course, National Centre for Research Methods Working Paper 04/13.

Thompson, B. (2007) Airport retailing in the UK, *Journal of Retail & Leisure Property*, *6*(3), 203–211.

Thompson, C., Clarke, G.P., Clarke, M., Stillwell, J.C.H. (2012) Modelling the future opportunities for discount retailing in the UK, *International Review of Retail Distribution and Consumer Research*, *22*(2), 143–170.

Timmermans, H. J. P. (1981) Multiattribute shopping models and ridge regression analysis, *Environment and Planning A*, *13*(1), 43–56.

Tomintz, M., Clarke, G.P. Alfadhli, N. (2015) Location-allocation modelling, in Brunsdon, C. and Singleton, A. (eds) *Geocomputation: a practical primer*, Sage, London, 185–198.

Trading Economics (2014) United Kingdom GDP Constant Prices, www.tradingeconomics.com/united-kingdom/gdp-constant-prices.

US Census Bureau (2014) Quarterly Retail e-Commerce Sales, 3rd Quarter 2014, US Census Bureau News, US Department of Commerce, Washington, DC.

Vance, S.S., Scott, R.V. (1994) *Wal-Mart: a history of Sam Walton's retail phenomenon*, Twayne, New York.

Vickerman, R.W. (1974) Accessibility, attraction, and potential: a review of some concepts and their use in determining mobility, *Environment and Planning A*, *6*(6), 675–691.

Vickers, D. (2006) Introducing geodemographics and area classification: multi-level integrated classifications based on the 2001 Census, PhD thesis, University of Leeds.

Vickers, D., Rees, P. (2007) Creating the UK National Statistics 2001 output area classification, *Journal of the Royal Statistical Society: Series A (Statistics in Society)*, *170*(2), 379–403.

Voas, D., Williamson, P. (2000) The scale of dissimilarity: concepts, measurement, and an application to socio-economic variation across England and Wales, *Transactions of the Institute of British Geographers*, *25*, 465–481.

Washington Post (2014) Target says up to 70 million more customers were hit by December data breach, *Washington Post*, 10 January.

Washington State Department of Transportation (undated) Population Growth in Relation to the State's Counties, Washington State Department of

Transportation, Olympia, www.wsdot.wa.gov/planning/wtp/datalibrary/population/PopGrowthCounty.htm (accessed 20 November 2016).

Weber, A. (1909) *Theory of the location of industries*, University of Chicago Press, Chicago, IL.

Weltevreden, J. (2007) Substitution or complementarity? How the internet changes city centre shopping, *Journal of Retailing and Consumer Services*, *14*, 192–207.

WHSmith (2014) *History of WHSmith*, www.whsmithplc.co.uk (accessed 20 November 2016).

Whiteaker, A. (2011) Little Waitrose roll-out to start in London, *Retail Gazette Online*, 4 July, www.retailgazette.co.uk/articles/32344-little-waitrose-rollout-to-start-in-london (accessed 20 November 2016).

Whittemore, A.H., Smart, M.J. (2016) Mapping gay and lesbian neighborhoods using home advertisements: change and continuity in the Dallas-Fort Worth Metropolitan Statistical Area over three decades, *Environment and Planning A*, *48*(1), 192–210.

Wilcocks, L., Sauer, C. (2000) (eds) *Moving to e-business: the ultimate practical guide to effective e-business*, Random House, London.

Willetts, D. (2013). *Eight great technologies*, Policy Exchange, London. www.gov.uk/government/speeches/eight-great-technologies.

Williams, H.C.W., Kim, S. (1990) Location-spatial interaction models: 3 competition between organizations, *Environment and Planning A*, *22*, 1281–190.

Williamson, P., Birkin, M., Rees, P. H. (1998) The estimation of population microdata by using data from small area statistics and samples of anonymised records, *Environment and Planning A*, *30*(5), 785–816.

Wilson, A.G. (1970) *Entropy in urban and regional modelling*, Pion, London.

Wilson, A.G. (1974) *Urban and regional models in geography and planning*, Wiley, London.

Wilson-Jeanselme, M., Reynolds, J. (2006) Competing for the online grocery customer: the UK experience, in Kornum, N. and Bjerre, M. (eds) *Grocery e-commerce: consumer behaviour and business strategies*, EE publishing, Cheltenham.

Witeck Communications (2015) America's LGBT 2014 Buying Power Estimated at $884 Billion, www.witeck.com/pressreleases/americas-lgbt-2014-buying-power-estimated-at-884-billion/ (accessed 20 November 2016).

Wood, S., Browne, S. (2006) Convenience store sales forecasting: art before science?, *European Retail Digest*, *15*, 15–18.

Wood, S., Browne, S. (2007) Convenience store location planning and forecasting: a practical research agenda, *International Journal of Retail and Distribution Management*, *35*(4), 235–254.

Wood, S., Lowe, M., Wrigley, N. (2006) Life after PPG6 – recent UK food retailer responses to planning regulation tightening, *International Review of Retail, Distribution and Consumer Research*, *16*(1), 23–41.

Wood, S., McCarthy, D. (2014) The UK food retail 'race for space' and market saturation: a contemporary review, *The International Review of Retail, Distribution and Consumer Research*, *24*(2), 121–144.

Wood, S., Reynolds, J. (2011) The intra-firm context of retail expansion planning, *Environment and Planning A*, *43*(10), 2468–2491.

Wood, S., Reynolds, J. (2012) Managing communities and managing knowledge: strategic decision making and store network investment within retail multinationals, *Journal of Economic Geography*, *12*(2), 539–565.

Wortmann, M. (2004) Aldi and the German model: structural change in German grocery retailing and the success of grocery discounters, *Competition and Change*, *8*(4), 425–441.

Wrigley, N. (1985) Categorical data methods and discrete choice modelling in spatial analysis: some directions for the 1980s, in Nijkamp, P., Leitner, H., Wrigley, N. (eds) *Measuring the unmeasurable*, Martinus Nijhoff, The Hague.

Wrigley, N. (1987) The concentration of capital in UK grocery retailing, *Environment and Planning A*, *19*, 1283–1288.

Wrigley, N. (1991) Is the 'golden age' of British grocery retailing at a watershed?, *Environment and Planning A*, *24*, 1537–1544.

Wrigley, N. (1994) After the store wars: towards a new area of competition in UK food retailing, *Journal of Retailing and Consumer Services*, *1*(1), 5–20.

Wrigley, N. (1998) Understanding store development programmes in post-property-crisis UK food retailing, *Environment and Planning A*, *30*(1), 15–35.

Wrigley, N. (2000) The globalization of retail capital: themes for economic geography, *The Oxford handbook of economic geography*, 292–313.

Wrigley, N. (2002) 'Food deserts' in British cities: policy context and research priorities, *Urban Studies*, *39*(11), 2029–2040.

Wrigley, N., Branson, J., Murdock, A., Clarke, G.P. (2009) The missing regional dimension to the competition commission's findings on entry and exit of small and specialist stores in British high streets, *Environment and Planning A*, *41*, 2063–2085.

Wrigley, N., Clarke, G.P., Hall, S., Purell, L. (2017) To what extent a 'polish grocer' effect? ethnicity and competition in UK urban high streets, (forthcoming).

Wrigley, N., Coe, N.M., Currah, A. (2005) Globalizing retail: conceptualizing the distribution-based transnational corporation (TNC), *Progress in Human Geography*, *29*(4), 437–457.

Wrigley, N., Currah, A. (2003) The stresses of retail internationalization: lessons from Royal Ahold's experience in Latin America, *The International Review of Retail, Distribution and Consumer Research*, *13*(3), 221–243.

Wrigley, N., Dolega, L. (2011) Resilience, fragility, and adaptation: new evidence on the performance of UK high streets during global economic crisis and its policy implications, *Environment and Planning A*, *43* (10), 2337–2363.

Yu, B., Yang, Z., Cheng, C. (2007) Optimizing the distribution of shopping centers with parallel genetic algorithm, *Engineering Applications of Artificial Intelligence*, *20*(2), 215–223.

Zwass, V. (1996) Electronic commerce: structures and issues, *International Journal of Electronic Commerce*, *1*(1), 3–23.

Index

For Product Safety Concerns and Information please contact our EU representative GPSR@taylorandfrancis.com Taylor & Francis Verlag GmbH, Kaufingerstraße 24, 80331 München, Germany

Batch number: 10726532

Printed by Printforce, the Netherlands